The Lost Cities of Colorado

Laurel Michele Wickersheim
and **Rawlene LeBaron**

HERITAGE BOOKS
2009

HERITAGE BOOKS
AN IMPRINT OF HERITAGE BOOKS, INC.

Books, CDs, and more—Worldwide

For our listing of thousands of titles see our website
at
www.HeritageBooks.com

Published 2009 by
HERITAGE BOOKS, INC.
Publishing Division
100 Railroad Ave. #104
Westminster, Maryland 21157

Copyright © 2002 Laurel Michele Wickersheim
and Rawlene LeBaron

Other books by the authors:

*Colorado on the Eve of Statehood: An Edited Business Directory
of the Pioneers Who Built the Centennial State*

Mine Owners and Mines of the Colorado Gold Rush

All rights reserved. No part of this book may be reproduced or transmitted in any form or by any means, electronic or mechanical, including photocopying, recording or by any information storage and retrieval system without written permission from the author, except for the inclusion of brief quotations in a review.

International Standard Book Numbers
Paperbound: 978-0-7884-2190-7
Clothbound: 978-0-7884-8117-8

TO

IVA B. HOSKINS ANDERSON

CONTENTS

Boulder County	Caribou	1
	Frances (Camp Frances)	17
	Hessie	23
Chaffee County	Alpine	25
	Babcock	30
	Beaver City	31
	Cashe Creek (Cash Creek)	32
	Hancock	35
	Harvard City	37
	Hortense	38
	Iron City	39
	Maysville	40
	Monarch (Camp Monarch, Chaffee City)	42
	Rockdale	44
	Romley (Murphy's Switch)	45
	St. Elmo (Forest City)	47
	Shavano (Clifton)	51
	Stonewall	53
	Turret	54
	Vicksburg	57
	Winfield	59
Clear Creek County	94 (Ninety-Four)	61
	Alice (Yorktown)	62
	Bonito	65
	Freeland (Trail Creek Camp)	66
	Lamartine	70
	North Empire (Upper Empire)	76
	Red Elephant	77
	Silver Creek (Daileyville, Chinn City)	78
	Spanish Bar	80
	Waldorf	81
	Yankee (Yankee Hill)	82
Costilla County	Russell (Sangre de Cristo, Placer)	83
	Isle	85
	Querida (Bassick City, Basssickville)	86

	Rosita (Brown's Spring)	88
Eagle County	Fulford (Nolan's Creek Camp, Camp Fulford)	94
	Holy Cross City	98
Gilpin County	American City	101
	Apex	102
	Baltimore	105
	East Portal	106
	Nevadaville (Bald Mountain, Nevada City)	107
	Nugget	109
	Tolland (Mammoth)	111
	Wide Awake	112
Gunnison County	Alpine Station	113
	Baldwin	114
	Bowerman	115
	Crystal	117
	Dorchester	120
	Floresta (Ruby, Ruby-Anthracite)	121
	Gothic	123
	Hillerton	130
	Irwin	133
	Marble	136
	North Star (Lakes Camp)	138
	Ohio City (Eagle City, Gold Creek)	141
	Scofield	143
Hinsdale County	Burrows Park	145
	Capitol City (Galena City)	146
	Carson	148
	Henson	150
	Sherman	152
	Whitecross	154
Lake County	Adelaide	156
	Brumley (Bromley Station, Bromley)	157
	Everette (Halfway House, Seiden's House)	158
	Leadville	159
	Malta (Swill Town)	175

	Oro City and New Oro City (Agassiz, Kelly's Diggings, Poverty Flats, Boughtown, Shaptown, California Gulch)	176
	Saint Kevins (Sowbelly Gulch, Amity)	178
	Stringtown	179
	Stumptown	180
	Tabor City (Tabor, Taylor City, Halfway House, Chalk Creek Ranch	181
	Twin Lakes and Dayton	182
La Plata County	La Plata City	184
	Parrott City	185
Mineral County	Bachelor (Teller)	187
	Sunnyside	191
Ouray County	Guston	192
	Ironton (Copper Glen)	194
	Red Mountain Town	196
	Sneffels	200
Park County	Alma	203
	Buckskin Joe	205
	Como	209
	Fairplay (South Park City, Platte City, Fairplay Diggings)	210
	Hamilton	214
	Horseshoe (Doran, East Leadville)	215
	Jefferson	216
	Leavick	217
	London Junction and Alma Station	218
	Park City (Mosquito)	219
	Tarryall	220
	Webster	221
Rio Grande County	Jasper (Cornwall)	223
	Summitville	224
Saguache County	Bonanza (Bonanza City)	227
	Iris	230
	Liberty	232

San Juan County	Animas Forks	233
	Eureka	241
	Gladstone	243
	Mineral Point (Mineral City)	245
San Miguel County	Alta	248
	Ames	249
	Pandora (Newport)	250
	Placerville	251
	Tomboy (Savage Basin Camp)	253
Summit County	Argentine (Decatur, Rathbone)	255
	Carbonateville	258
	Chihuahua	259
	Conger (Conger Camp)	261
	Delaware Flats (Delaware City, Preston, Braddock)	262
	Dyersville	264
	Kokomo	267
	Lincoln (Paige City, Lincoln City)	274
	Masontown	278
	Parkville (Park City)	279
	Preston	280
	Rexford	281
	Robinson (Camp Robinson, Ten Mile)	282
	St. John (Coleyville, Sts. Johns)	286
	Swan	288
	Swandyke	291
	Swanville	292
	Tiger	293
Teller County	Altman (Midway)	294
	Anaconda (Mound City, Squaw Gulch, Barry)	296
	Cameron (Gassy, Grassy)	297
	Elkton (Beacon Hill, Eclipse, Arequa)	298
	Gillett	300
	Goldfield and Independence	301
	Stratton (Winfield)	303
Photographs		unnumbered pages

References	305
Index	307

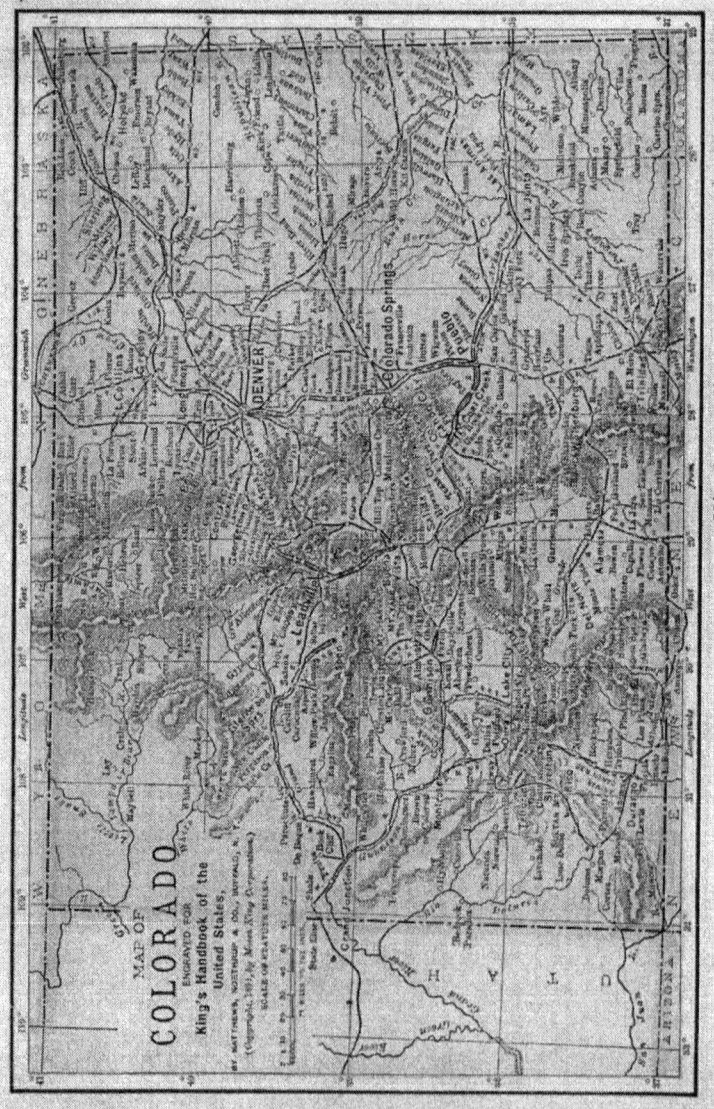

CARIBOU
Boulder County

Directions: Caribou is five miles northwest of Nederland, 22 miles west of Boulder, and 60 miles northwest of Denver. From the northwest edge of Nederland, take County Road 128 for five miles. This road will take you directly to the Caribou town site. Just before the town site, the Caribou Silver Mine is on the left side of the road, marked by a large sign.

After driving for a few more minutes, you will pass the remains of two stone buildings on your right. Continue the short distance to the top of the hill and park. This is Caribou Hill, the town site. On the left side of the road at the base of the hill was Caribou Flats. Any car can make this trip, but if you want to go farther on this road to Rainbow Lakes, you should have a four-wheel drive vehicle with clearance. This is not a dangerous road, and camping is allowed.

The History of Clear Creek and Boulder Valleys, Colorado, published in 1880 by O.L. Baskin & Co. reported Sam Conger's discovery of the great Caribou silver veins. In the fall of 1860 Sam Conger was elk hunting about a mile east of Klondike Mountain, when he noticed some black rocks, later identified as lode blossom, on the ground. He did not think this black rock was valuable and did not give it another thought until he recognized the rocks again, about eight years later. Sam Conger was in Larimer City when he noticed a broken box of Comstock ore that had been shipped from Nevada on the Union Pacific Railroad. This silver ore was the same black rock he had seen while elk hunting in the mountains of Colorado. That fall, he returned to Colorado, and in the spring he found the blossom rock that would lead him to the rich vein of silver he sought. He called his discovery The Conger, and his campsite was the future site of Caribou. That spring, he took five partners: his friends William Martin and George Lytle, George Lytle's son-in-law Samuel Mishler, John H. Pickel, and Hugh McCammon. They soon discovered the source of the silver ore, staked their claim, and George Lytle named their lode The Caribou after the Caribou gold diggings of Canada.

Sam Conger traded his share of the Caribou Mine, which would eventually gross $8,000,000, for 100% of the rights to the Poorman Mine, a parallel vein that also proved to be a good producer.

That fall the men cut a trail to Black Hawk to transport a load of Caribou ore to Professor Nathaniel Hill's Mill and Smelting Works. They made it just before winter. The ore proved to be as rich as they dreamed. The men spent the winter mining the silver ore and stockpiling it for the next spring.

By June 1870, word spread about the rich silver strike in Caribou, and hundreds of men came to the area. They pitched tents, built cabins, brought their families, and searched for silver. Some of the discoveries that summer were the Idaho, the No-Name, the Seven-Thirty, the Sovereign People, the Spencer, and the Trojan.

The town of Caribou grew up, named for the mine.

From the spring of 1871 until August of 1872, Collier & Hall published *The Caribou Post*, the community's weekly newspaper. The reporter was A. Bixby who lived in Caribou. Only seven issues are preserved on microfilm in the Denver Public Library archives. These important few surviving copies sketch the details of everyday life in a little boomtown that no longer exists.

Local and Miscellaneous

This morning, at the shaft house of the Caribou Company, we found them raising ore from the splendid sheet of silver-bearing mineral a few days ago uncovered in the end of the mine. It was a fine site to see, the heavy blocks of sulphurets coming to the surface from the depth of one hundred feet. Some of these chunks were two feet in thickness. Heavy pieces lay on the dump, estimated to contain not less than a dollar per pound.

In the spacious ore house of the Caribou mine - Breed & Cutter - we saw today, a sack containing 225 pounds of fine black and grey sulphurets, at least the rate of five thousand dollars per ton. This and many sacks of ore of the value of five hundred dollars per ton, or near that value, more or less. This is from the drift. The water is out of the mine. It will be left to drain a few days when work will be actively resumed in the bottom, where the best pay lies.

Mr. Goldring has forwarded a box of Caribou ore to London. It is probable the east half of the lode will soon pass into English hands. The success which would undoubtedly reward such an enterprise must tend to greatly enhance the value of Grand Island mines.

Leo Donnelly's store on Caribou street, is a favorite outfitting point for all in want of supplies. Mr. Donnelly keeps posted in the matter of prices, and puts goods down as low as they can possibly be afforded. He has on hand a full supply of everything in his line, for which see advertisement in the *Post*.

Walter H. Smith, of Boulder, met Mr. Spottswood here the other day to receive the formal transfer of the stock and mail contract of the Caribou Stage Co. on the line from Erie to Caribou, via Boulder City. Mr. Smith has long done the principal part of the staging, express business and mail carrying for Boulder county, and is known by everybody as attentive and prompt in business, and as one who takes pleasure in making traveling pleasant to all of his passengers. His coaches leave Boulder every morning, meeting the cars at Erie at noon, returning in the afternoon. Monday, Wednesday, and Friday, leave Boulder for Caribou, returning on alternate days. Mr. Lee Smith is the gentlemanly driver on the mountain route. By no other way can persons desirous of visiting the mountains go from the valley to the snowy range by so direct a route, or on such easy grades, or witness such grandeur of scenery by the way.

The Planter House, W. O. Logue, proprietors now about completed. It is a large, thoroughly built, commodious house, two and a half stories high. On the second floor, which opens on Caribou street, are the sitting rooms, ladies' parlor, and four large airy sleeping rooms. On the lower floor, which opens to the south, is a dining room 14 by 28, convenient cooking rooms, and back of these the family rooms. The upper story is partitioned off into sleeping apartments. The cellar is the best we have seen in the mountains. Mr. Logue has a full house of boarders and strangers, as the order and style in which things will always insure. Among the arrivals at this house by last evening's coach, are Mr. Morgan and wife from Denver.

The Caribou Post
Saturday, 8 July 1871

Local and Miscellaneous

Wednesday afternoon the clouds did us the grace of a few drops of rain.

W. O. Logue & Son have one of the newly discovered silver veins on the north slope of Caribou Hill.

The attention of the farmers through the county is directed to the price current of Howes & Breath.

The Boulder News comes with an amusing account of Independence Day at Caribou.

Mr. Goldring, from England, and D. H. Moffat, cashier First Bank, Denver, were in town yesterday.

The American Flag is the name given to their rich discovery on the Caribou Hill, by Lesher & Burnam.

Our friend, Pete Thompson, is pleasing himself with the prospect of a fortune out of the Ten-Forty. "What`s to hinder?"- Not the lack of silver ore surely.

The Saratoga lode, near the Shamrock, on the Idaho hill, is the property of Gov. McCook, T. J. Oyler and others, who express a high opinion of its value.

The Grand View is attracting much attention. Two shafts are worked. The size of the crevice and the quality of the ore raised, mark it as a first-class mine.

The boys are not yet done speaking in praise of the dance at the Cardinal House a few evenings since. It is regarded as the most festive occasion of the season.

Dr. Hopkins, recently from southern Illinois, has seven or eight prospecting. He is bound to see the lode to be his fortune before leaving. His faith and his works insure his success.

Work is steadily progressing on the Potosi, and the constant accumulations of smelting ore on the dump, indicate that the industrious owners see the road to prosperity.

An extensive fire is raging in the timber at the head of Stewart Gulch, and on Silver Creek. Already, as we learn, it has consumed 1,000 cords of wood in pile.

The Peabody is the name given to the new silver discovery of McNeal & Reynolds on the Caribou hill. T. J. Oyler has bought a fourth interest, and the work of development is prosecuted continuously.

The Morning Star is a lode of brilliant promise. The owners, Shea, Smith, & Kimball, are sinking the main shaft without cessation. Solid mineral is appearing in the bottom, and much valuable ore is raised.

We are very choice of a specimen of silver galena, presented us by Mr. Allum, the assay value of which is at the rate of twenty-eight thousand dollars per ton. It cuts like pure silver, and must be very near that precious point.

The Idaho is about to shine in more than former splendor. This lode is coming out of the dead ground with a two and a half foot crevice. Mr. Westover informs us that the ore appears better than any raised last season.

The fire which broke out near Tucker's saw mill last Wednesday, has already consumed five cabins in the timber, subjecting the owners to much loss in provisions, clothing, etc. and is now sweeping up very near the town on the east and south.

A fire started Sunday morning on Leavenworth Mountain, in the timber near Leavenworth Gulch. Several cabins have been already reduced to ash. The first owned by Mose Hawk, and occupied by the Cowles Brothers, and Glass, was burned with all of its contents, consisting of several hundred dollars worth of miners' supplies, and clothing. The fire was spreading towards the National Mill on

Sunday evening. The origin seems to be generally charged to a party who were preparing to make charcoal. They had been burning the brush off. They believed the fire out, but a breeze sprang up that started it again. We need rain very much. Corn is curing on the stalk. Vegetables are literally drying up. Fires are liable to break out in the timber at any moment. We have not had so dry a summer by the recollection of the oldest inhabitants here.

The Caribou Mine

Yesterday, by the politeness of Mr. Martin, Superintendent of the Caribou mine, and in the company with our dear friend Mr. Allum, the foreman in the mine, we made a full exploration of all of its shafts and galleries, taking accurate measurements, and upon which we make the following notes.

Situation

The Caribou location is on the crown of the hill, about one-fourth of a mile westward from the town of Caribou, and in full view of the snowy range, whose peaks begin their ascent about one mile further [sic] west.

A pay-chute, or chimney, as it is called, was indicated, where the main shaft now is, by a mass of blossom overflowing from the swelling of the vein to the surface at that point. From the centre of this discovery shaft, which is the dividing line, the property of the old Caribou company extends on the vein eastward 700 feet, and that of Messrs. Breed & Cutter, the same distance west.

General Characteristics of the Lode

The course of the lode is nearly east and west, varying, we believe, one degree to the south of west, and one to the north of east. Near the surface the lode dips considerably to the north, but gradually assumes a vertical position, and at the bottom of the main shaft, now a hundred and eighty-five feet deep, is turning a little southward. Both walls are composed of hard, dark colored granite, standing not far apart, the intervening space being filled with a larger proportion of paying ore, and a smaller proportion of gangne, than we have ever before observed. Heavy stulls set in hitches in these firm walls, make the security of the mine perfect. The ores are chiefly sulphurets, brittle silver, and silver glance, the latter very plentiful, and giving the ore extraordinary riches. Very little galena.

Buildings

The main house over the mine is 130 feet in length, by 40 feet in breadth. This gives ample space for the assortment and storage of ores, engine room, warming room and storehouse. The latter contains a well-filled powder magazine, and all goods required in the prosecution of the work. The outside buildings consist of large houses over the east and west shafts, hay barn and blacksmith shop. These are all framed to withstand the winter winds.

The First Level

This was run each way from the main shaft, westward 50, and eastward about 30 feet. The rich ore commenced in a thin sheet at the surface, but soon thickened to several feet, and continued so down to this level; all above which was worked last fall, with results that gave to the mine its world wide celebrity.

The Second Level

This is 60 feet below the first-extending west 187 feet, and 100 feet into the east half of the mine. Perhaps half of the ground between the two levels is worked out. We were shown a square of unworked ground, 40 feet on each side, and cut all around, exhibiting a pay crevice averaging two feet in thickness, and the ore worth in the mass $250 per ton. A rich deposit had just been struck in the workings near by, which yielded 21 sacks, 50 pounds to the sack, of high grade ores, worth some thousands per ton. All the best ore is carefully sacked by the workmen in the mine--all the coarser rock is raised and sorted above with much pains.

Below the Second Level

The ore crevice continues about the same as above for 40 feet downwards, at which point a scam of ore, which left the main vein at the first level, comes in again, and the pay widens to six feet. It soon narrows to four feet, and continues that width to the bottom of the shaft, 85 feet below the second level, and 185 feet from the surface. Here Mr. Allum struck a pick into the solid mineral, and the first chip which flew we took to assayer Kearsing, and obtained a certificate of 410 ounces and 4 grains of silver per ton, -- coin value

$533. Mr. Allum estimates the average value of the whole width of the crevice at 380 ounces of silver per ton. Thus it is seen that the ore increases both in quantity and richness as depth is attained. And depth will be carried as far down as human skill can go in search of the precious metals.

Manner of Working the Mine

The object of the foreman now, is to open the mine in such a manner that, if required, a large number of men can be worked with the same advantage as a few. So that if 100 tons of ore per day was demanded, it could be raised at the same cost per ton as a small quantity. And Mr. Allum is accomplishing this admirably. Only a few men are now worked in the mine, yet they are sending forward considerable ore to Hicks. They keep three heavy teams running constantly to Black Hawk, but a portion of their loading is now of ore previously raised. The mine is already in condition to supply the company's large mill, now building at Dayton, four miles below, and which it is expected will be completed in September.

Shafts

West 50 feet from the main shaft is another 40 feet deep, connecting with the upper level, and this is connected with the second level by a winze. West 200 feet from the main shaft is yet another 90 feet deep, the workmen in which, within a few days, will meet the men who are driving the lower level. The 500 feet, surface measurement, still further west on this property, is yet unbroken ground. East from the main shaft 100 feet, another shaft is worked by the old company. This is 117 feet deep. In this half of the mine their first level connects with the second level of the west half. At the bottom of this shaft, 71 feet below the first level, they are now just beginning their second level, each way. This company have eastward 600 feet of unoccupied ground. This east half of the mine is in the richness and abundance of its ores, the counterpart of the west. At the bottom of this east shaft we asked the probable value of the ore being broken in large blocks from a vein two feet in thickness. We are told that it was second-class ore, and would yield about $400 per ton. "If this is your second-class ore," we inquired, "what is your first-class?" We were answered that the first-class consisted of smaller deposits of fine sulphurets and silver glance, with [worth] from $1,000 to $5,000 per ton. What shall be said of a mine whose second-class ore produces $400 per ton.

In General

These two properties, whose division line cuts the main shaft in the centre, properly constitute but one mine, and both together but one location, on a mother vein. After being once on pay, there seems to be no barren places in the ground yet explored. In pinched spots the ore is compressed to the thickness of six or eight inches, and this yields about $500 per ton. Such places are worked single-handed. One man picks out in a day the value of $50. We observed no such narrow places below the second level, but . .[illegible] . thickness of the ore-vein We see here a continuous sheet of ore, compressed into the compass of four feet in thickness, carrying as much metallic value as forty feet in width of vein in the Comstock mine. The amount of ore sold from the time of the purchase, in September, 1870, to June 1st, 1871, was 428 tons, averaging $176.58 per ton, showing a gross value of $75,576.55. Besides this, several tons of fourth class ore, assaying about $50 per ton, were piled up to await the completion of the mill before spoken of.

It is too early to ask for the figures of this year's operations; nor will the mine be made to show anything like its bullion producing capacity until the companies have their own mills. From the park on the north side of the hill, Messrs Breed & Cutter have secured a tunnel site, which will cut the vein at considerable depth. Were a tunnel, on the vein, to be commenced far enough east to run under Caribou hill at the depth of two thousand feet, it would not be an undertaking greater than has been found necessary in European silver mining. In the brief and imperfect notes submitted, we have sought accuracy of statement which should characterize all that is written concerning this great mine, probably the most valuable of any now worked in any part of the world.

The Caribou Post
Saturday, 22 July 1871

A Serious Accident

Occurred in the Caribou mine on Monday last the 31st July. One of the workmen, a Mr. Jenkerson, descended the main shaft by ladder to the lower platform, from which point the remaining distance is made by the bucket, into which Mr. J. stepped. The well-understood signal for lowering is two strokes of the bell, for raising one stroke. Either Mr. J. by mistake, gave the signal for raising or made the second stroke so light that the engineer did not hear it. So the bucket began to rise, taking Mr. J. upwards. At some distance above, he attempted to reach the bell-rope to make the signal for lowering, which threw the bucket out of its place, and becoming entangled in a brace the man was thrown out, and fell to the bottom, a long distance, 40 feet or more, and was very badly bruised, having a small bone of the right arm broken. He is fortunately under the care of a skillful physician, Dr. Hopkins, and unless internal injuries prevent, will recover.

A Leap in the Dark
Yesterday the Longmont party visited the Caribou mine. They descended the west shaft to the platform, which is within about 25 feet from the bottom. It was pretty dark below. The conductor of the party stepped on a bench cut in the rock on the side of the shaft a few feet below, preparatory to making the descent by holding on to the rope. One of the party, supposing he (the guide) was standing on the bottom, made a jump and brought up 25 feet below on a pile of rock just thrown out by a shot. His legs and hips were badly cut, but no serious injury done.

The Caribou Post
Saturday, 5 August 1871

Grand Island men will accede to anything reasonable in the way of county aid for a railroad to Caribou.

During the past week many refreshing showers have fallen upon the mountain farms. Grain and vegetables have been planted to a much greater extent than usual, and promise well.

Jenkerson, who was badly hurt in the Caribou mine, a few days ago, is rapidly recovering. The Longmonter [resident of Longmont, Colorado] who made the famous leap in the dark, has gone home to wash his bruises, and bind up his wounds.

The Caribou Post
Saturday, 12 August 1871

At and About Caribou

The town is assuming a much more lively appearance with the increased prosecution of mining which is now going on. The extreme depth of the snow which fell last winter, and immense quantity of water which was thereby entailed on the mines, made the season very late. The large increase of the force in the Caribou mine, and the activity on the other property is being felt. There are now about 50 men employed in the Caribou, and we are informed that it is the intention of Messrs. Breed and Cutter to increase the force to two hundred. A contract has been let for running a tunnel 750 feet in length to cut this mine at a depth of about 350 feet. Work on this tunnel is to be prosecuted day and night, and the whole work completed at as early a day as practicable. Messrs. Breed and Cutter are also working with the idea of purchasing the Mount Vernon lode on the summit of Caribou hill. The work on this lode is prosecuted for the sake of development. Some pay ore is being raised and a wide, main crevice is shown.

Havens & Bottleson are working the Sherman lode, north of the Caribou. They are drifting on the vein and taking out five tons per day of ore, but little inferior to the Caribou ore.

Sam Conger has leased the Poorman, also north of the Caribou, for eighteen months and is taking out some good ore.

A contract has been let on the Morning Star, and six men are engaged in developing this property, which is proving quite valuable.

The Silver Point, north of the Idaho, owned by Moore, Schweeder & Livernash, is being worked with very satisfactory results. Some working capital was raised in Denver for the development of this mine. Some extraordinary rich ore has been found, specimens of which have assayed as high as $10,000 to $15,000.

The Berger Brothers are working the 5-20, which is now twenty-five feet deep, has a pay streak ten inches in thickness, and looks quite promising.

The Boulder County, Coral, Mammoth, and Sovereign People at Cardinal are being worked with very satisfactory results.

The Caribou Post
Saturday, 17 August 1871

Evidence of Prosperity
Everywhere the new buildings bear the character of permanence – in favorable contrast with those of former years. This is especially noticeable at Caribou, where buildings are of the most substantial kind- more the appearance of a permanent town than any mining town we have ever known. Almost without exception the shafts on the new mineral locations are made up with heavy timbers.

The Caribou Post
Saturday, 2 September 1871

The town of Caribou was nicknamed "the place where the winds were born." Any season of the year, the winds blowing across Caribou can almost carry you away. Snowstorms are brutal. Whiteout conditions and massive 25-foot snowdrifts could blanket the town during winter months. It was common to have to come and go through second-story windows.

Caribou had become a prosperous boomtown.

When General Grant visited Central City during the summer of 1872, silver bricks from the Caribou mine were laid from the carriage stop to the door of the Teller House, a 150-room hotel that had been opened earlier that year. The silver bricks were pulled up long ago, but the Teller House is still standing and was renovated as a casino.

For 1878, *The Colorado State Business Directory* reported:

CARIBOU

Mining town in Boulder County, Grand Island Mining District; tri-weekly mails and express; money order postoffice; stages to Central, Nederland and Boulder. Population, 500. The principal lodes are the Caribou, No-Name, Sherman, Poorman, Idaho, Virginia, Grand View, Morning Star, Wabash, Native Silver, Monitor, Potosi, Missouri Valley, Pomeroy, Croesus, Homestake, Ten-Forty, Great Republic, Seven-Thirty, Norwood. These mines are producing $50,000 per month. The principal tunnels are the Caribou, Idaho, Airline, Summit, Red Cross. Distances: To Boulder, 22 miles; Central [Central City], 20; Hot Springs of Middle Park, 60; Denver, 50.

WM. H. WOODS,
Bureau of Mines and Mining Interests,
Notary Public, Justice of the Peace, Conveyancer and Collector, Caribou, Boulder County, Colorado. (Lock Box 20.) Careful and compete attention and supervision given to the interests of absent investors in the mines of this vicinity.

CARIBOU DIRECTORY
Postmaster-S. B. HARTER.
Town Clerk-P. J. WERLEY.
Town Trustees-J. J. Murphy, Wm. Scott, C. A. Sherwood, John
 Simmons, Joseph Lloyd.
Police Judge-LEO DONNELLY.
Churches. Methodists and Presbyterians occupy the school-house.
Societies. Caribou Lodge, No.22, I.O.O.F.

BUSINESS DIRECTORY
Assayers. COLE, GEORGE. Livernash, Adolphus.
Bureau of Mines. WOODS, W. H.
Butcher. Murphy, Jos. J.
Fruit. Reimer, J. M.
General Merchandise. HERZINGER & HARTER. SCOTT
 BROTHERS.
Hotels. CARIBOU HOUSE, S. Richards. PLANTERS` HOUSE, John
 Pierce. HERMAN HOUSE, M. Gorman.
Livery Stable. COSGROVE, JOHN.

Notaries. Donnelly, Leo. WOODS, W. H.
Physicians. DAVIS, J. R. H. MANN, WILLIAM J. Stock, John.
Restaurant. NEWELL, SAMUEL.
Saloons. BESHORE & GRAHAM. Dodson, H. R. Jordie, John. Newell, Samuel. SEARS, WERLEY & CO.
Stage Lines. SMITH & CO., (Walt. H. and Leon), tri-weekly to Nederland and Boulder. Smith & Co., (Walt. H. and Leon), tri-weekly to Central City.

Caribou's residents were originally an optimistic, enthusiastic people, and the little town was bustling and prosperous. That was, until September 1879. One Sunday fall morning, the town woke up to a red haze in the western sky. Forest fires had blazed across the area for a couple of weeks. It was reported that someone had been careless with a campfire. The town had running water, but unfortunately the water main had broken at the lower end of town in Caribou Flats below the Sherman House. The buildings over the Caribou mine were destroyed, along with the eastern half of the town. After the fire, the town rebuilt itself, but the population dropped considerably.

Caribou flourished in the 1880's, and the population peaked at 3,000. After the silver crash of 1893, the few mines that were open and running struggled to survive.

By 1896, Caribou had fallen on hard times. *The Colorado Business Directory for 1867* reported that the Caribou was now a mining camp – only 20 years earlier it was a bustling mining town with a shining future -- and its population had shrunk to only 100 people. The town had only a handful of listings.

CARIBOU
Mining camp in Boulder County, 22 miles west of Boulder. Six stages per week to Boulder and Central City. Population 100.

HINMAM, MRS. S. E, gen mdse and postmistress
Huggins, Chas, stage line
Thorp, Louis, hotel
White, James, justice peace

It was the day after Christmas, 26 December 1899, when another fire engulfed Caribou, destroying most of the town. One of the few surviving structures was Sherman House, a huge, three-story, U-shaped white hotel, on the south corner of Potosi Street and the west side of Sherman Street. Along with the Sherman House, the church and another hotel were saved from the fire. The residents fled to Nederland. They came back only to look at what was left of their homes. Many, knowing they had lost everything, never returned.

Caribou was struggling for survival, and the remaining residents could not afford to rebuild the town. A railroad had never been built all the way to Caribou, making it that much harder to rebuild. When transporting ore out of Caribou, the men used wagons strapped to mule and horse teams. The men would take the "Coon Trail," the road to Cardinal, three miles to the southeast. From Cardinal, Nederland is only two more miles to the southeast.

Cardinal was a little suburb of Caribou. It was a little town that got its start after some of Caribou's citizens attempted to clean up the red light district, which stretched along Potosi Street and lower Idaho Street. These citizens chased the prostitutes, gamblers, transients, and alcoholics out of town, until finally the refugees simply set up camp outside of Caribou. They built houses, saloons, and gambling halls, and before long the men of Caribou began visiting the little camp. This makeshift camp became the town of Cardinal. Ironically, there are still a few people living in Cardinal to this day. . . unlike Caribou, whose only residents lie in the cemetery.

From the top of Caribou Hill, a dirt road heads directly east. Follow this to the top of the hill, where the cemetery is. The writing on the headstones has worn off, and some of the graves were never marked. As the years went by, some of the coffins disintegrated, creating a shallow depression in the soil. Death records show that many of the children buried in the cemetery died during two deadly epidemics of scarlet fever and diphtheria, diseases that we now routinely immunize our children against.

Almost nothing is left of the town site of Caribou. The partial remains of a log cabin mark the Caribou hillside above Caribou Flats, just past the remains of two stone buildings on the right side of the road at the entrance of Caribou. The town site has become a beautiful mountain field and hillside. It is almost unimaginable that a city once existed in this spot; there are no roads, no foundations, just nothing. The authors combed the town site with a metal detector and found nothing except a few square nails. The entire Caribou town site was built on top of an iron dyke, which is basically a sheet of iron drawing electrical storms in the spring and summer.

Just past Caribou Hill is the marked trail to the Rainbow Lakes area of Caribou, which is in the middle of the Indian Peaks Wilderness and Roosevelt National Forest. At this point a four-wheel drive vehicle is necessary. The U.S. Forest Service usually closes the trail during the winter, but it is open from Memorial Day to Labor Day. Excellent campsites can be found here.

Ruins of the old Potosi boarding house in Caribou. This boarding house was built to accommodate miners working at the Potosi mine, one of the principal lodes in the area. One can imagine the noise and energy of the miners who lived there over a hundred years ago.

FRANCES (Camp Frances)
Boulder County

Directions: From Ward, go south on the Peak-to-Peak Highway (Highway 72). The site of Frances is about a mile south of present-day Ward, on the left side of the Highway. It was located at the mouth of a gulch.

Ward was a mining camp named for Calvin W. Ward, the prospector who discovered the Ward lode in 1860. Ward was once a successful mining town with a population of about 4,000, but today only about 200 people live there.

The little town of Frances grew up just south of Ward. Originally called Camp Frances, it was simply Frances when it peaked in the 1890's with a few hundred residents. It had grown up as a little mining town, and its main employer was the Big Five Corporation.

The Ward newspaper, *The Ward Miner*, covered the daily news and social events in Frances. Annual subscriptions to *The Ward Miner*, under the editorship of H. T. Haines, cost $1.50 a year, payable in advance, or $2.00 if not paid in advance. A sampling of some of the columns printed in 1899 draw the life of the community of Frances.

FRANCES.
Frances, Colo.
Jan. 11, 1899.

G. W. Kennedy went to Denver Sunday to meet his wife's sister, who is coming to visit them for an indefinite time.

Thad Adams went to Boulder Tuesday on Business.

Professors Lake and Loitus and Mr. Huffman were visitors in camp several days last week.

The Sunday school is reviving. S. B. Merrill was elected Superintendent; Miss Duncan, Assistant Superintendent, and Arthur Adams, Secretary, last Sunday.

Mrs. S. R. Kinsman came to visit her husband on Wednesday.

Miss Book returned Sunday to resume school on Monday, but owing to such an extensive amount of travel while the trains are snow bound the very cautious school board was afraid someone might bring small pox into camp, and so not to permit school to open until the pupils were vaccinated, and had a certificate to that effect.

The Christian Endeavor meeting next Sunday afternoon will be led by Miss Duncan. Special music has been prepared. All are cordially invited to attend.

The Ward Miner
Friday, 13 January 1899

FRANCES
Frances, Colo.
Jan.19, 1899

We Frances people are reveling in clover—have a train once a week on an average.

W. H. Blair has accepted a position on the C. & N. railroad.

J. C. Garmon has struck it rich in his lead on the Buck Horn and is liable not to be back for several weeks.

The Ward Miner
Friday, 20 January 1899

A Stormy Week
The past week has been a record breaker for cold and storms. Monday was the coldest day, the thermometer registered 26 below zero at 7 a.m. and the mercury hovering close to the bottom of the glass all day. Tuesday was another cold day, the lowest showing being 18 degrees below. It has stormed more or less every day of the week, and generally more than less. At times the wind has blown

terrifically, and with the snow, made outdoor work almost impossible.

Local Briefs

Every sort of vehicle can be seen in daily use here except an automobile, and may be one of them will be coming along soon.

Tuesday next is St. Valentine`s day. We trust our fool friends will send us the average number of remembrances of the fact that they are not all dead yet.

The Ward Miner
Friday, 10 February 1899

Local Briefs

S.B. Merrill, Civil Engineer of the Big Five is quite seriously ill at Frances.

Mines and Mills

Mining at Frances is still held in check because of the huge drifts of snow that cover the railroad tracks between that camp and Puzzler.

The Columbia Mines company, to operate the Columbia and the other mines in Ward, filed articles of incorporation last week, with a capitalization of $3,000,000. The concern is related to the Big Five at Frances and the nine directors are: William P. Daniels, N.C. Merrill, H.A. Hendrickson, C.G. Tallman, H.A. True, J. P. Loftus, and Robert Binder. The 500 incorporators represent most of the states in the Union.

The Ward Miner
Friday, 10 March 1899

FRANCES.
Frances, Colo.
April 13, 1899

Mining will be resumed in the Big Five properties next Monday, and everybody is happy. The camp is emerging from the snow and little patches of terra firma are visible here and there. Those who are "killing time" in the valley are returning and the camp is filling up with life and activity, and cheerfulness is apparent all around.

A. E. Gill, the medical student who is filling the position of meat cutter and salesman in Miss Duncan's store, is becoming quite popular. Mr. Gill expects to return to the state university next fall and complete the course, his absence now being to secure the means to allow him to do so.

John Maier and family have been quite ill with grippe.

We have become spiritualists among our people and séances have become quite common of late.

P. R. Hinkle is expecting a lot of finishing lumber with which to complete his house.

The house warming dance given at O'Tooli's new hostelery last Thursday night was a great success. In fact, it was the social event of the season and attended by everyone.

J. H. Welsh, the Adit machine man, has been spending the week in Denver. M. B. Carter, Lafe Latta, and Geo. Glass were among the returning miners from Denver one day during the week.

The Ward Miner
Friday, 14 April 1899

FRANCES.
Frances, Colo.,
May 17, 1899
S. B. Merrill is now at the Homeopathic hospital in Denver, and his health was in such precarious condition this week that his mother was summoned from Frances to his bedside.

James Watson and Jerry Harding, two of Frances most respected young men, left for Denver Wednesday, where positions were offered them.

A baby daughter arrived last Saturday to bless the home of Mr. and Mrs. John Conklin.

Rev. Harger found a large audience last Sunday afternoon to welcome his genial self to camp again.

S. R. Kinsman expects to leave for Cripple Creek in a short while, believing that there dame fortune is more easily wooed.

The school term closes next Friday with appropriate exercises. Miss Book will depart for Wyoming a few days later.

A dance was given at Martin Shaft's residence last Saturday night, which was largely attended. The affair was given as a farewell testimonial to Messrs. Watson and Harding.

Miss Duncan last week purchased the Baxter house, which she now offers for sale or rent, consideration, $300.

Thad and Ed. Adams have a big contract for putting in wood for the Wall Street mines.

The people of Frances are making preparations for a reception to be given at the Adams home, Saturday evening, May 27, in honor of Miss Book, who had so successfully managed the school at this place during the past year. Miss Book is an accomplished lady and a competent teacher, and besides her systematic management of the school she has been an active church and Sunday school worker and a leading light in the social life of the camp. Hosts of admiring friends will regret her departure very deeply. The people of Ward are cordially invited to this reception.

The Ward Miner
Friday, 19 May 1899

Area mines were the Adit, the Columbia, the Dew Drop, and the Ni Wot. It is not known exactly when Frances was founded, but Frances appears to have existed as a community since at least 1867 when the Ni Wot mine was worked:

The Old Ni Wot in Line

Great as is the past record of the old Ni Wot mine for production, a glance into its upper workings to-day will satisfy the most skeptical that the most interesting history of the famous old property is yet to be made. Old timers say that the old Ni Wot has produced a million and a half of gold and talk entertainingly for hours of the days when it was the only mine working in camp; of shipping its ore 35 miles to Black Hawk, at an expense of $80 per ton for hauling and treating. Even at this great cost it paid to work the property. But no one can give a satisfactory reason why work ever suspended on the mine, save the litigation, the inevitable result of mismanagement of inexperienced and impractical mining men, intervened. Seven months ago Wm Tobie and Israel Benson, men who worked on the Ni Wot as long ago as 1867, entered into an arrangement with the Big Five concern to work the upper workings of the property on a percentage plan. They took into partnership with them on this scheme George Kerkett and the senior and junior Jack Holman.

The Ward Miner
19 May 1899

Winters were brutal, stopping daily activities for weeks at a time. Until the spring of 1901, nothing very bad had ever happened in Frances. Snow had built up in the mountains and on the railroad tracks at Frances, and the passenger train was scheduled to depart. The engineer and the crew were determined to stay on schedule. A huge drift blocked the train's way, so before the train hooked up the passenger coach, the engine attempted to charge the drift. That's when a massive avalanche roared down the mountain, carrying the train with it to the bottom of the gulch, killing the entire crew aboard. The town was in shock, some people moved away. Mining continued, but over the next decade everyone gradually moved away.

The site of Frances is visible from Highway 72, but nothing is left of the little community. The only reliable information on Frances is found in very old newspapers.

HESSIE
Boulder County

Directions: Any car can make this trip. From Nederland, take County Road 130 west for five miles. After three miles, you will pass through the old town of Eldora. After driving the five miles, park at the parking lot and follow the signs to the town site.

The population of Hessie peaked in the 1890's, with about 300 residents in the area. By 1900, only 16 people lived in "Hessie Town," according to the 1900 Federal Census:

- G.P. Herman, the grocer, and his wife. He was born in Ohio; she was from Canada.
- P.J. Manihulin [?], a carpenter, his wife and two children, who had moved to Colorado from Maine.
- J.B. Balch and his daughter Nellie, both school teachers, from Indiana
- M.A. Cheney, a 53-year-old widower from Georgia who made a living by "keeping boarders."
- W.G. Nesbit, a 62-year-old carpenter, and his wife Laura G., who had both been born in Indiana, and their 20-year-old son, Dave [sic] G., an unemployed day laborer, who had been born in Iowa as his family had migrated west.
- E.A. Witten, his wife Anna, and his brother, N. A. Witten. They had all migrated from Missouri. The brothers were employed as day laborers.
- L. S. Allnott [?], a 43-year-old gold miner from New York. He was married, but his wife was not living with him. Like so many prospectors who rushed to Colorado, he probably left his wife and children behind until he made his fortune.

Hessie was located at the junction of North Boulder Creek and South Boulder Creek, in a thick forest of pine trees. It had been named after the postmaster's wife, Hessie. The two neighboring camps were Grand Island, which was a few miles up the north fork of the creek, and Lost Lake, which was about a mile to the southwest. Residents of Hessie had high hopes for their town, but the mines played out before the turn of the century. The residents left for more prosperous towns. The few cabins that remain have been sheltered by the trees.

After parking your vehicle follow the path that you will see. You will walk alongside the creek for a bit, you will soon cross a bridge, which is in good condition. A sign put up by the Forest Service welcomes hikers and campers to the Indian Peaks Wilderness area. Continue following the trail for a short distance, and you will come upon the last few cabins.

ALPINE
Chaffee County

Directions: Alpine is about 16 miles from Buena Vista. From Buena Vista, go south on U.S. Highway 285, and turn right going west on State Highway 162.

Alpine sprang up in 1877, when the first cabin was built here. It was a little mining town, and in 1881 the railroad came to town, and turned Alpine into a railroad construction town, too. In 1880, the Federal Census reported that the population was 503, but a large transient population, estimated from 1,000 to 5,000, was not counted. Alpine had many stores, beautiful wooden sidewalks, 23 saloons, 3 banks, a big dancehall, and a newspaper. The early newspaper, called *The True Fissure*, was moved to St. Elmo in 1880. In 1883 the town's sampling works employed around 40 men. Ore from the Tilden Mine was processed for shipment by wagon to Pueblo.

According to legend, Mt. Princeton's 14,197-foot chalk cliffs, now called Nathrop's chalk cliffs, are the hiding place of a stolen Ute treasure. These cliffs are west of Nathrop, overlooking the town site of Alpine. In the early 1800's, a party of Spanish explorers surprised a village of Ute Indians. Most of the men of the village were away on a hunting trip, making it easy for the Spanish explorers to rob and pillage the defenseless Indian village. The Spaniards left with the gold and valuables that the Indians had accumulated. When the Indian men returned, they found destruction and death, and some of them set out to pursue the Spaniards. When the Spanish raiders heard the Indians after them, they hid the treasure cache in the chalk cliffs and tried to flee. The avenging Utes caught and killed most of the Spaniards. Only a few escaped, but they never returned to unearth their stolen treasure. For over a hundred years, people have searched the area, using antique maps and prospectors' journals, metal detectors, and aerial scans, all without success. Still hidden, somewhere in the cliffs, are two leather mule packs of stolen Ute treasure.

Newspaper articles of the time optimistically report the growth and early prospects of the area.

Mountain Items
The Alpine Lode is creating quite an excitement recently, especially among the property holders of the lode. From a prospect taken from near the surface of Faul & Clark`s claim, a yield of 30 cents to the

pan was obtained and the crevice well defined. Should it continue to pay as well as it promises, it will rank among the first in the country.

The Rocky Mountain News
Wednesday, 10 February 1864

Alpine
Interesting Mining Notes from that Busy Camp

In 1873 an old California prospector by the name of Royal, not meeting with success on the Gunnison, conceived the idea of following up Quartz Creek, a tributary of the last mentioned river, to the summit of the continental divide, thence to the Arkansas Valley. His route lay over high mountains to the eastern slope where rises Chalk Creek, a tributary of the Arkansas River. Descending the last mentioned stream he soon discovered heavy mineral stain on the granite walls of the canon, which encouraged him to spend several months in prospecting the region, now known as the Alpine camp, and the staking of a large number of claims.

In the following spring, the road being opened from Canon City to the mouth of Chalk creek, a considerable number of prospectors visited the camp, among whom was Mr. Chapman, of Kansas City, one of the first settlers in the present town of Alpine, and one of the most respected citizens.

At this time, 1874, the rush was for the San Juan, while Rosita was just coming into notice and Leadville's vast treasure vaults yet concealed from mortal vision. The subsequent discovery of mineral in the great carbonate camp drew all the world thither, leaving the small camps to a sickly existence and ephemeral growth.

But later the refluent wave of prosperity that Leadville brought to the State has reached the smaller camps, and Alpine, an outlying camp, has awoke within the last twelve months to the possibilities that are now within her reach, and promises to become the most promising mining camp of Chaffee county.

A road will soon be built, as it is under contract, between Buena Vista and the Continental, following Chalk Creek canon, through Alpine, thence up the south fork of the creek near to the summit, where a tunnel 1,800 feet long will complete the road across the mountains to the head waters of Quartz Creek, a tributary of the Gunnison, down which the road will follow to Pitkin City, four miles distant from Alpine. The work in question is all under contract. The road is to be finished to Alpine June 1 and through the tunnel July 1. It is but recently that Alpine has attracted the attention of Leadville and Eastern capitalists. Considering the comparatively small amount of labor expended in mining here, the results are very encouraging. Within a few months some very valuable veins have been worked, containing large ore bodies and rich in high-grade mineral. If the camp were provided with reduction works or facilities for shipping ore by rail the product of the camp would surprise those not familiar with its resources.

Prof. Falco, a mining expert and one of the best posted men in mining affairs of the camp, states that within thirty days the output of ore, if there were a market for it, would reach 200 tons per day, and 400 tons before the end of the season. Of the mines that will at once yield rich returns to their owners, when the railroad is finished, are the Brickensten mines on the Murphy group, Black Hawk, Iron Chest, Tilden, Jessie, Munroe, and the Lake View group. These mines alone, it is claimed, can furnish sufficient ore to run steadily four smelters, each of sixty tons capacity. A notable fact with reference to the quality of the ores is the large proportion of argentiferous galena, which are rich in lead. If railroad connection with Leadville were completed, your smelters could obtain a large supply of these fluxing ores as are now in such great demand in your camp.

Referring to shipments of ore made from Alpine last season to Denver, Canon City, Leadville, and Newark, the returns as shown by the books of Mr. Falco were 85, 90, 99, 130, 147, 175, 287, and 308 oz. Silver. During the present season some very large developments of mining properties will be made by tunnels. Among these mentioned the Chrysolite, which when finished will cut 16 veins; the Brickensten, 11 veins; Lake View, 24 veins; Pettibone, 6 veins;

besides the Harrison and Commonwealth, which will develop valuable mining properties. Leadville is already well represented here. Edwin Harrison, Governor Tabor, Senator Harris Jr., C. H. Clemens, E. H. Hazelton and Messrs. Jackson and Davis, have recently secured valuable mining properties here, which, as soon as the railroad is finished, to Alpine, promise the fortunate owners large returns.

The Carbonate Weekly Chronicle
Leadville
28 February 1880

The Colorado State Business Directory for 1878 lists the following information on Alpine:

ALPINE
Mining town in Lake County. Population, 300 to 500 in the mining season.

ALPINE DIRECTORY
Postmaster — J.E. McClure.
Justice of the Peace — Chas. R. Fitch.

BUSINESS DIRECTORY
Blacksmith. Duchat, Ed.
General Merchandise. Evans, R. W. & Bros
Groceries. Wilsey & Ray.
Hotels. Alpine, Brittenstein.
Liquors. Leahy & Scott.
Miners Supplies. McClure, J. E.
Physician. Wright, A. E.
Saloon. Merriam, George D. & Co.
Smelting Works. Kansas City Smelting Co.
Concentration Works. Cambell Bros.
Mines. Tilden, Riggens, Hively, Whipple, Continental, Hortense, Little Mattie, Anna, Little Virginia, Minnetonka, Clinton, Shoo Fly, Evening Star, Mammoth.
Tunnels. Chrysolite Co., Commonwealth Co., Hudson & Fish, Grizzly, Shoo Fly, Black Hawk, Mary Murphy, Tilden, Topeka, Yankee Boy.

In 1881, when the railroad stretched to St. Elmo, people gradually moved there.

Today, nothing is left of Alpine but the site and a few summer cabins.

According to legend, Mt. Princeton's 14,197-foot chalk cliffs, now called Nathrop's chalk cliffs, are the hiding place of a stolen Ute treasure. These cliffs are west of Nathrop, overlooking the town site of Alpine.

BABCOCK
Chaffee County

Directions: Leaving Garfield northwest on Forest Route 230, which follows the Middle Fork River, drive for four miles. This will take you to the site of Babcock. Or, if you are coming from the site of Hancock, and if you have a good jeep, go to the far end of the town site, where you will find two trails. The trail on the right leads to Alpine Station. Do not take this. Take the trail that goes left; it is marked with a sign saying to Hancock Lakes. You will pass the unmarked site of the old town of Stonewall, which died ten years before other towns in the area. All that remains is the Stonewall Mine. Keep following the trail. It is about two miles from Hancock to Hancock Lakes. After the Lakes, you will cross over Chalk Creek Pass at an elevation of 12,000 feet, and then it is about another mile to the site of Babcock, which will be on your left.

Five mining camps sprouted in the Mount Antero area following a silver strike in the late 1870's. In addition to the silver, the slopes of Mount Antero conceal small pockets of aquamarine, hexagonal blue beryl crystals. Babcock was the largest of five camps in the area; the others were Green's Gulch, Jennings, Hartz, and Foosel. The silver at Babcock was very pure, but because it was isolated and transportation was very difficult, little is known about this camp. It may have been named for Colonel Babcock, owner of the Mountain Chief Mine. Little remains of Babcock, and a visitor would be lucky to identify a foundation.

BEAVER CITY
Chaffee County

Directions: Beaver City is nine miles from Granite. From Granite, go south on U.S. Highway 24 for about three miles until you reach County Road 390. County Road 390 traces the historic Colorado Trail and Clear Creek. Turn right on County Road 390 and continue about six miles until you reach the site of Beaver City.

A camp was established after prospectors found some good float in Clear Creek in the early 1880's. This was essentially a tent city with few buildings. The camp did not last long. Residents moved farther up the creek when Winfield and Vicksburg were built.

CASHE CREEK   (Cash Creek)
Chaffee County

Directions: From Granite, follow Cashe Creek west upstream for 2 ½ miles. The town site is on an unmarked road next to the Cashe Creek road.

In 1860 placer gold was discovered along Cashe Creek. In response, many little mining camps popped up along the creek: Cashe Creek, Gold Run, Oregon Creek, Ritchie's Patch, and other small ones. Winters were cold, the gulch was steep, but the gold along the creek lured prospectors. The town of Cashe Creek was founded in 1866, and townspeople built the first post office in Chaffee County. One source says the population was 200, while another says 3,000.

Abe Lee and a party of gold seekers from Georgia had come to Cashe Creek before discovering California Gulch to the north. Horace and Augusta Tabor stayed at Cashe Creek briefly before moving on to Leadville. The Tabors, like Abe Lee and his group, were unsuccessful here. There was a lot of fine gold here, but placer mining demanded patience that many of the prospectors did not have. It was a long, tedious process extracting gold that was so heavily laden with the black sands.

Cashe Creek did well from 1860 to 1867, yielding almost a quarter of a million dollars in metals. Eventually the mining started to slow, and the residents drifted to Granite or north to Leadville. The only traces of a once-bustling community are some tailings piles, some well-hidden foundations, and a cemetery.

Nearby Lost Canyon, about three and a half miles down the gulch from Granite, was named for the many prospectors in the 1860's who had gotten lost there. It is known for a lost trove of huge gold nuggets. In 1860, a party of miners coming from Gunnison got lost. While trying to find their way back, they thought they might as well check the area for gold. When they found a few nuggets the size of chicken eggs, they decided to stay as long as possible. In just a few weeks they had found $60,000 in gold. It was already fall, and the threat of early winter snows forced them to return to camp. They planned to return in the spring. But, in the spring, they could not find the site that yielded the huge nuggets. No one knows if the spot was ever found, or if it was covered over by an avalanche.

The following *Rocky Mountain News* article sketches the winter dangers and hardships that threatened the prospectors:

FROM THE ARKANSAS MINES
CASH CREEK, Jan. 1, 1864

EDITOR NEWS: A happy New Year to you. May kind fortune ever smile upon you, is the prayer of your humble servant, the subscriber.

The old year expired with us in a fearful death struggle, the wild conflict of the elements at times was truly appalling. The wind blew a perfect tornado and mountains of snow seemed moving through the air. Old Boreas done a grand day`s work in filling the mountain ravines. I have never witnessed a more terribly stormy day than yesterday. The weather was not so intensely disagreeable by the drifting snow and searching wind.

The snow drifts in the gulches are from ten to twenty feet deep in places, at least one-third more snow has fallen up to the present time, than fell the whole of last winter, and it is still snowing nearly every day, especially on the high ranges. Today the weather is more pleasant but by no means settled. Miners are delighted to see snow accumulating on the mountains, for it is a sure guarantee of a bountiful supply of water for the coming mining season. *Nihil desperandum* is our motto.

Everything seems to be working admirably in our favor. In the afternoon of the last day of last year, Mr. Robert Short, of California Gulch, arrived here from Fair Play with a number of jacks laden with flour and other articles of grub, making our hearts glad, for these supplies will carry us nearly through the winter. Are we not thankful for these favors? Yes, truly; are we. "Blessed are those whose ask for little, for but little shall they receive;" I don`t know whether this is scripture or not, any one who has the curiosity may ascertain.

We feel now that winter will pass more rapidly and pleasantly. You may wish to know how we spend our time, of which there is such a surplus, well we eat two meals per day, with a lunch when we can get it between times. One hour in the morning and evening is spent in procuring and preparing wood. The balance of daylight is generally spent within the bosom of our families, except occasionally when we make raids upon the sage hens and the mountain sheep that abound in the vicinity. The game usually comes off harmless, but not so with the luckless Nimrods; they often returning with frozen extremities. The evenings are spent in paying friendly calls up on each other, reading, writing, and playing sundry games. Don't understand that we gamble, for we are too moral for such wickedness. The noble game of chess is most indulged in. There are two champions, Mr. Wilburn Christianson and myself. We have but little advantage over each other, consequently the tournaments are interesting to ourselves only, except now and then when one another of us gets blind, then the one that can see well gets all the games. Now don't understand that blindness here refers to the use of ardent spirits, for Cashe Creek has been strictly temperate for near three month. Water, coffee, and tea, when we can get the two last, form our only beverages. We once in a while have a game of euchre or whist at the residence of Mr. Boyd. These games only occur when strangers come to town, otherwise we cannot raise a quorum. But enough; more anon.

The Rocky Mountain News
Tuesday, 26 January 1864

HANCOCK
Chaffee County

Directions: From Buena Vista, go south on State Highway 24 to Nathrop, then turn right on State Highway 162. From Nathrop drive west, and follow the road as it turns south. After 22 miles you will reach Hancock. The road turns into a dirt road, but if you are cautious and the weather is good, you should not need a four-wheel drive vehicle.

The town of Hancock was formed in 1880 by miners working on the Hancock Placer claim, although it was never officially incorporated. In 1880 the railroad already had a stop here. The tracks are now gone, and a road has been built over the old rail bed. On the left side of the road, hidden by underbrush, there is still the gray railroad grade gravel, and a few scattered pieces of railroad ties are imbedded in the ground.

A little bit of mining was done in the area. When driving into the town site, to your left is where the Flora Bell Mine stood. The Flora Bell and the Allie Bell had their loading bins next to the track below the town. On the end of town to the left was the Stonewall Mine, the best producer in the town. It produced mostly galena, a soft lead sulfite crystal formation, from 1879 until 1915. Galena is the main ore of lead, although silver is a common byproduct.

When driving into the town site on the main street, to the left appears to be a rocky meadow at the base of a mountain, with traces of the lower few logs of a cabin. A U. S. Forest Service sign identifies this last standing structure in Hancock; it was the saloon. Archived photographs from the late 1800's from the Western History Museum show the current meadow filled with stores, businesses, homes, and cabins.

Hancock was more of a railroad town than a mining town. Railroad workers who built the Alpine Tunnel lived here. The town was only two miles from the lower eastern end of the tunnel. The tunnel was 17 feet high, 14 feet wide, and 1,805 feet long. By the time the tunnel was finished November 1881, it had cost $120,000 and took 300 men to build it. The tracks were threatened by frequent rockslides in the summer and deadly snow slides in the winter. One particularly bad snow slide buried 13 people, including children, beneath 30 feet in snow. Railroad employees had to shovel the tracks to keep the train from derailing. At least twice an avalanche thundered down the mountain taking train cars with it. From the Alpine Tunnel entrance, all the way down to the town of Hancock, the grade was the steepest in the state. At times the trains would start from the beginning of the tunnel at full speed, and plow through the snow for as long as they could, just like a mole. If the train stopped while it was still buried by the snow, the engineer had to climb out and shovel off the smoke stack so the smoke would not fill the train compartment and suffocate them.

Mr. E. Wilber, roadmaster of the South Park Railroad, hooked up a flatcar behind a passenger train, and took it to the top of the pass. There, he hopped on the flatcar, unhooked it from the train, and rode the flatcar two miles down the hill. His only control during this thrill ride was with a handbrake. Mark Twain was visiting the area when he heard of this thrill ride and insisted that he, too, might try it. He later claimed that the two-mile flatcar ride was the wildest and most exciting adventure he had ever had.

Hancock was a thriving railroad town in 1881. Main Street was a line of false fronted buildings: five stores, many saloons, a hotel, and a couple of sawmills. Main Street faced the railroad tracks. In 1881 the population was 200.

Tunnel maintenance was so expensive that in 1910 the tunnel fell victim to economics and was closed. The closing of the tunnel led to the demise of Hancock. The tracks had been ripped up long ago, leaving a scenic two-mile trail from Hancock to the Alpine tunnel. While closed to motorized vehicles, it is an enjoyable hike or bike ride.

Hancock was a thriving railroad town in 1881, with a busy Main Street lined with false fronted buildings. Today, the only remnant of this once boisterous town is the foundation of a saloon.

HARVARD CITY
Chaffee County

Directions: From Buena Vista, take County Road 306 west for seven miles. Harvard City is located at the confluence of Middle Cottonwood and South Cottonwood Creeks.

Prospectors had been placer mining in this area since 1860. County Road 306 from Buena Vista to Harvard City, and then to the Continental Divide was the Cottonwood Pass Toll Road. Harvard City was a busy supply center and convenient rest stop for travelers. There were quite a few saloons, a dancehall, and even a post office. Miners had been developing lode mines in the area in 1874. Many travelers passed through this area in the 1870's; most were headed to Aspen. Cottonwood Pass was treacherous in the winter, and many lives were lost. Several different sources had reported that in 1879, a group of men had made a large tunnel going through 90 feet of snow pack only to reach a cliff. At that point they got their ropes out and proceeded to lower their wagons, supplies, and then themselves successfully 700 feet off the edge of a cliff. In 1881 Independence Pass was opened, providing an alternate route to Aspen. This new pass was easier and much safer; travelers no longer used Cottonwood Pass. The miners in the area moved on to places like Leadville, where rich strikes were made everyday. Buena Vista had now become the new supply center.

Harvard City was abandoned, and today there is very little to mark the little mining community.

HORTENSE
Chaffee County

Directions: From Buena Vista, go south on U.S. Highway 285 to Nathrop. At Nathrop take County Road 162 right for five miles. Then turn right, and stay on the road that stays the closest to the chalk cliffs for a mile. This is the site of Hortense.

The small town of Hortense was started as a mining camp in 1871, although the post office did not open until 1879. Hortense and Mount Princeton Hot Springs were so close to each other that it was impossible to tell where one town started and the other ended. Miners had cabins scattered throughout the area. In 1881 the railroad came through town. By 1887 there were 100 people in the community. Across the street from the Hortense train station was Mount Princeton Hot Springs Hotel.

The miners who lived in Hortense, worked at the Hortense Mine, high up the slopes of Mount Princeton, 12,000 feet high. This was the first silver mine in the area, founded by Captain Merrian in 1871. The path up to the mine was narrow, treacherous, and very difficult. Mine manager Eugene Teats later built a safer road that wound up the mountain for five miles.

IRON CITY
Chaffee County

Directions: From Nathrop on U.S. Highway 285 go right, which is west on County Road 162. Drive for about 13 miles. It is after the town of Alpine, and about a mile before the town of St. Elmo. There will be a small sign on the right side of the road, saying Iron City, take a right, and follow the road a few hundred feet. The town site has been converted into a campsite.

Iron City, a small town next to St. Elmo, peaked in the 1880's. The town's economy revolved around a smelter, built in 1880 to treat ore from area mines, and an electric generating plant that was built in 1912. The Iron City Power Plant was built to provide electric power for placer mining operations in Tin Cup, a little over 15 miles away, for the Tin Cup Gold & Dredging Company in Gunnison. The company chose the site at Iron City for their power plant because they could pipe water from the St. Elmo Dam, 7,000 feet away, and make a 262-foot waterfall to power the plant's generators. The water was then returned into Chalk Creek. Although the power plant experienced financial problems from the beginning, it was finally closed in 1917, when a World War I scrap metal drive confiscated the metal from the machinery, including the metal that was in the wood stove pipe, which piped the water from St. Elmo and created the waterfall.

The railroad soon came through the area, but it did not stop at Iron City. All of the local miners went to St. Elmo to ship their ore out of town, rather than using Iron City's smelter. The townspeople left. Iron City's Reservoir washed out what was left of the town.

The remains of the smelter foundation are to the right of the gateway to the town. The Iron City Cemetery is a few hundred yards farther and to the left. At the unlocked gate of the cemetery is a sign naming everyone who had been laid to rest there and how they died. White marble and oval wooden nameless gravestones set in the fenced cemetery surrounded by trees. A little farther up the dirt road, to the left is the little power plant building, the last remaining building in Iron City.

MAYSVILLE
Chaffee County

Directions: From Buena Vista, take U.S. Highway 285 south until you reach U.S. 50. At U.S. 50 turn right, which is west, for about six miles. Maysville is exactly six miles before Garfield, 228 miles from Denver, and 12 miles west of Salida.

Maysville was a mining town that sprang up in July of 1879 on the Feathers' Ranch, owned by Mr. Amasa Feathers. After silver was discovered on his ranch, Feathers divided his ranch into lots and sold them for $75 each. Within days, the area was filled with tents. The town was named after General William Marshall's Kentucky hometown of Maysville. General Marshall had recently discovered Marshall Pass. The mining here was good, two smelters were built to treat the ore, and then in 1881 the railroad came to town.

The Monarch Pass Toll Road was finished in 1880, offering a route over the Continental Divide to the town of White Pine. At the edge of Maysville was a tollgate, and that road went over the divide to the town of Black Sage. Another toll road built in 1880 went from Maysville to the town of Shavano. This road followed the north fork of the Arkansas River for ten miles.

In 1882 *The Colorado Business Directory* reported a population of 600, with the Paymaster Mine as a top silver producer, and the Erie Con. Mining and Smelting Company as one of the town's main companies. Many hotels, one of which was named the Hughes' House, stores, and saloons filled the business district. The Denver and Rio Grande Railway had a branch that ran through town. Good cattle country surrounds Maysville.

The two local newspapers, the *South Arkansas Miner* and *The Maysville Chronicle*, competed with each other, both bragging about having the highest circulation in the county.

In 1885, *Crofutt's Grip-Sack Guide of Colorado* listed the population at 150. George A. Crofutt was born in Connecticut in 1827 and moved West, where he became famous as the writer of some of the most important guidebooks of the nineteenth century American West. Some of his titles included *Great Trans-continental Railroad Guide*, first published in 1869; *Crofutt's Overland Tourist and Pacific Coast Guide*, in various volumes and revisions, first published in 1878; and *Crofutt's Overland Tours*, first published in 1888. His books were eagerly bought by a nation that had become caught up in the gold rush and adventure of the American West.

The Daily Budget, the Ouray newspaper, briefly mentioned that a fire on 6 July 1888 "destroyed the most valuable block in town."

The town rebuilt after the fire, but the Silver Crash of 1893 emptied the town.

By 1896 *The Colorado State Business Directory* reported that Maysville did not have any businesses:

MAYSVILLE
Old mining town in Chaffee county, on the D & R G R R, 6 miles from Poncha Springs. No business. No postoffice.

Nothing remains of this once prosperous silver boomtown.

MONARCH (Camp Monarch, Chaffee City)
Chaffee County

Directions: Monarch is on U.S. Highway 50, about 14 miles west of Poncha Springs. From Garfield, Monarch is located southwest on U.S. Highway 50, three more miles away in a narrow gulch.

In 1878 Nicholas Creede was a young prospector who set out with a horse, wagon, and grubstaked supplies to the Monarch Pass area in search of silver. He was grubstaked (provided with supplies in exchange for a percentage of a future strike) by a man named Boone. Later that year the young Creede discovered the Monarch Mine and the Little Charm. The Monarch Mine held thin erratic veins of silver that were graded at 200 ounces of silver for every ton. A 3,000-foot tramway system was built to transport ore from the Monarch Mine to the Denver & Rio Grande railroad railhead. Denver is 237 miles away. The next mine to be discovered was the Madonna, another top producer. There are conflicting reports, but some say that Creede also discovered the Madonna.

Prospectors came to the area by the hundreds, filling the gulch with tents. The men called their tent city Camp Monarch, for the mine. On May 15, 1879, the town was officially incorporated by Miller, Boone, and J.G. Evan, and renamed Chaffee City. People did not use this name, and it was officially changed to Monarch in 1884.

By 1880 Monarch was booming, and prospectors were everywhere, overturning rocks and digging holes in the surrounding mountains. The population peaked at 3,000. Well over 100 houses were built, along with three stores, three hotels, three saloons, and three assay offices. A Mining Exchange Building was built to exhibit ore samples from the different claims in the area. Soon the Eclipse and the Silent Friend mines were discovered. Ore had to be shipped to Canon City 100 miles away until 1881, when Maysville finally attracted a railroad to their town. Then, they had to haul the ore about six miles northeast to the Maysville railhead.

In 1883 the new owner of the Madonna Mine, Anton Eilers, was able to extend the railroad into Monarch to his mine. The Madonna Mine employed 300 miners with a day shift and a graveyard shift. Some men worked in the marble quarries outside of town. More men worked in lumbering and construction, helping to build the boomtown.

In 1893, the silver crash hit. The mines shut down, and most of the residents left.

By 1896, only 25 people lived in Monarch, according to *The Colorado State Business Directory* for that year

MONARCH

An old Mining camp in Chaffee county, on the D & R G R R. Population 25. Altitude 10,000.

Blatchford, G. E., postmaster and gen mdse
Heack, John, livery
Moore, D H, justice peace

The residents who remained in the dying town eventually tore down most of the buildings for firewood. The railroad stopped running. A new highway went over the pass. Then, avalanches destroyed the remnants of the town.

You might be able to find a few building remnants away from the road.

A big limestone quarry was opened by the Colorado Fuel and Iron Company next to the town site. An entire mountainside is terraced with tracks that lead to openings in the mountain. Limestone and zinc ores are packed onto the trains and transported to Salida.

Exactly a mile and a half below Monarch is the reported town site of Hartville. This little town had a restaurant, a saloon, and some cabins. It appears that nothing is left of Hartville.

An interesting four-wheel drive road at Garfield goes north to the old mines of Taylor Gulch.

ROCKDALE
Chaffee County

Directions: From Granite go south on U.S. Highway 24 for a few miles until you reach County Road 390, then turn right. Rockdale is 13 miles west of Granite, in the middle of Vicksburg and Winfield.

Rockdale was a small, short-lived mining town that sprang up in the early 1880's. It was located next to Clear Creek. The gulch is studded by many trees. If you are lucky, you might find what is left of a couple of cabins.

ROMLEY (Murphy's Switch)
Chaffee County

Directions: From Buena Vista, go south on U.S. Highway 285 down to Nathrop. Turn right, which is west, on County Road 162, and follow that for about 16 miles to St. Elmo. When you get to St. Elmo, on the left edge of town is Forest Route 295. It is a good dirt road. Take this for 2 ½ miles to the site of Romley.

Romley was discovered in the 1870's and called Murphy's Switch until 1897. It was a little mining camp that grew up close to the Mary Murphy Mine, as did St. Elmo and Hancock. The Mary Murphy Mine was named by a local Irish prospector who had been very sick, and cared for in Denver by a nurse named Mary Murphy. The mine was discovered in 1875. The men credited for the discovery were Dr. A. E. Wright and Mr. John Royal. They sold the mine for $75,000. The new owners were forced to sell the mine for $80,000 when their Kansas City Smelter in Alpine was not as successful as planned. In 1880 the Mary Murphy Mine started producing more gold and silver than any other mine in the district. Reportedly, 250 miners from Romley, Hancock, and St. Elmo worked at the same time in the mine. Mary Murphy produced over $60 million in silver, gold, iron, zinc, and lead.

An estimated 2,000 people lived in the Romley area at its peak. Romley is in a mountain meadow between St. Elmo and Hancock. In 1883 the railroad came into town, boosting the already prosperous economy. At the base of Pomeroy Mountain, a tramway system stretched for nearly 5,000 feet and ran 96 buckets, each one holding 200 pounds. The mine produced about 100 tons a day, the good ore assaying $125 a ton in gold and silver. The tramway ran from the fourth level of the mine to the railroad switch. The railroad station contained a mine office and a boardinghouse with accommodations for 40 men. Romley also had a boardinghouse at the timberline for the 60 miners.

No one knows why, but during 1909, all the buildings in town were painted fire engine red, and the doors and windows were trimmed in white.

That same year, an English company bought the mine and invested $800,000 in improvements. The mine still employed 250 miners until 1917, when it stopped producing. The owners leased the mine a few times, but without success.

When the Mary Murphy Mine closed down, it marked the end of Romley. The Pat Murphy and other small mines in the area simply could not save the town.

In 1926 the railroad came in and removed the tracks.

In the 1980's the town's buildings were bulldozed. The only traces of the little boomtown are the remnants of old cabins scattered about the hillsides.

Remnants of the Mary Murphy Mine in Romley, which was discovered in 1875. Reportedly, 250 miners from Romley, Hancock, and St. Elmo worked at the same time in the mine. Mary Murphy produced over $60 million in silver, gold, iron, zinc, and lead.

Long-abandoned aerial tramway to the Mary Murphy Mine in Romley.

ST. ELMO (Forest City)
Chaffee County

Directions: St. Elmo is 24 miles southwest of Buena Vista. From Buena Vista go south on U.S. Highway 285. At Nathrop, take County Road 162 west (right), for 16 miles and you will be at St. Elmo. The road to St. Elmo is a good one.

St. Elmo got its start as a mining town in 1870, when the town was still called Forest City, because of the thick beautiful pine trees that surrounded the town. In 1880 the citizens formally incorporated the town. According to *The Colorado State Business Directory* for the year 1882, the population for that year was 250, and the town had railroad service by the D & S P Railroad. When the post office department reviewed the papers for Forest City, they discovered that there was already a Forest City in California. The town chose a new name, St. Elmo.

St. Elmo is at the center of five gulches: Chalk Creek's North Fork, Chalk Creek's South Fork, Pomeroy, Poplar, and Grizzly. Two toll roads were at the upper edge of town: one went to Gunnison, and one went to Aspen. A constant flow of traffic went through town, as miners went from their mining camps on the other side of the Continental Divide. Freighters, merchants, travelers, and fortune hunters filled the streets of St. Elmo. From the 1880's and the 1890's the area population hovered around 2,000. Many of St. Elmo's residents were employed at the Mary Murphy Mine at nearby Romley.

In 1881 the railroad came into town; this was the last stop for the Denver, Southpark, and Pacific Railroad. Weary travelers could rest for the night in one of St. Elmo's five hotels. The largest was the Briscoe Block, the town's first two-story building. After the travelers had a chance to rest, and it was time to get back on the road, they could travel by stagecoach, horseback, by foot, or by wagon. Of the routes out of St. Elmo, the most heavily traveled was the Chalk Creek Toll Road, leading to the silver fields. Other routes were the Elk Mountain Toll Road or the Altman Pass Road. As soon as Independence Pass was finished, it eclipsed all other routes to become the most traveled pass. Denver is only 153 miles away, and Tin Cup is 15 miles away.

The first newspaper came out in August of 1880. It was originally called the *St. Elmo Rustler*, but after the first issue the publishers changed the name to the *St. Elmo Mountaineer*.

In 1882 the first school was built, the St. Elmo School. Occasionally Bishop Macheboeuf and Father Dyer would hold their church services there. Usually they held their services in St. Elmo's saloons, gambling halls, and bunkhouses.

St. Elmo was a prosperous little community. In 1888, *The Colorado Mining Directory* listed many employers: the Chrysolite Mining and Smelting Co., Harrison Tunnel Co., the Little Nell, Murphy M. Co., Mammoth Tunnel Co., and the Stanley.

The Denver Republican, a major Denver newspaper, reported that a fire destroyed most of the business district in 1890:

BIG FIRE AT ST. ELMO
A Terrible Blow To The Little Mountain Mining City
THE BUSINESS SECTION IN RUINS

Total destruction of the Clifton Hotel, Postoffice, and Hurleys Warehouse. Together with many other structures-Lack of water gives the flames a full swing among the light pine buildings- List of sufferers and their losses.

St. Elmo, Colo., April 19.-[Special.]- A very destructive fire occurred early this morning, in which nearly the entire business portion of the town was destroyed. As the town was devoid of either a water-works system or fire company, the efforts of the citizens were confined to saving stock, fixtures, and personal property.

The fire originated in the Clifton Hotel and after the destruction of several buildings on that side of the street jumped across to J. D. Criss's supply store and gutted the entire block on that side, with the exception of Whittenberger's grocery on one corner and Boyle's restaurant on the other.

All the burned buildings were composed of very pitchy material, and the flames oftentimes leaped into the air a distance of 50 or 75 feet, making it uncomfortably hot for the brave workers.

The Losses
The losses are enumerated as follows:
Clifton Hotel, furniture, $2,500; insurance, $750; owner, J.J. Stewart; building, $4,000; insurance, $500; owners, W.W. Hessey and Tom Stark.
Pat Hurley's warehouse and contents, $2,000; total loss.
L. Yockey, shoe store, $500; total loss.
John Rieson's bayery [bakery] and lodging house, building $2,000, contents $1,000: no insurance.
Office of the Colorado Mineral Belt Mining Company Building $600, contents $600, no insurance; owned by W.R. Logan & Co.
Two vacant buildings next to Boyle's restaurant, one owned by J. J. Stewart, $900; another by Dan Williams, $1,000.
Butcher shop owned by John Haenni, stock $1,000, building $1,000.
Post office and drugstore, contents $4,000. owned by F. W. Brush; partially covered.
Post office building, $1,000, with $500 insurance; owned by S. L. Merrill.
J. D. Criss' supply store building, $1,000, and $4,000 in stock; partially covered.
Two buildings owned by Baldwin and Savard, containing a saloon; buildings $1,800, stock $2,000.

The Denver Republican
Sunday Morning
20 April 1890

The general store and the post office shared the same building, and just before flames destroyed the building, the storekeeper ran into the post office and dragged out the mail. The town's drinkers and smokers were angry with the storekeeper for saving the mail and letting the alcohol and tobacco burn. They seriously, and vehemently, questioned the storekeeper's priorities. The town was rebuilt.

After the boom years, people started to move away until the town was almost abandoned. The town is a very well preserved historical landmark, and walking through town is a trip back in time. Today it is protected and maintained by a group of volunteers called "The Ghosts of St. Elmo." The Main Street is lined with stores and other structures, all in excellent condition. The Stark Bros. Store and Post Office is a well maintained, unpainted large wood building, and all the windows are intact, further protecting the interior of the structure. The St. Elmo Jail and Town Hall is a one-story, freshly-painted white building with bell tower.

Outside of town, are remnants of two box cars used as crew quarters by the Colorado and Southern Railroad. Crews lived in these box cars from 1908 until 1922. In 2001 these boxcars were added to the Colorado State Register of Historic Properties.

There are summer cabins in the area, and the area is a favorite site for camping, hiking, fishing, four-wheeling, or simply going back a hundred years to a different pace of life.

St. Elmo's Stark Bros. Store and Post Office is a well maintained, unpainted large wood building, and all the windows are intact, protected and maintained by a group of volunteers called "The Ghosts of St. Elmo."

Remnants of two boxcars outside of St. Elmo that had been used as crew quarters by the Colorado and Southern Railroad. Crews lived in these boxcars from 1908 until 1922. In 2001 these boxcars were added to the Colorado State Register of Historic Properties.

SHAVANO (Clifton)
Chaffee County

Directions: To go to Shavano, go to the town site of Maysville on U.S. Highway 50, and right after you pass Maysville, on the right side of the road take Forest Route 240. Follow this for about six miles, and you will be at the site of Shavano. Forest Route 240 follows the North Fork of the Arkansas River. The road is rough and steep, a jeep is recommended.

In 1879, gold was discovered, and a camp called Clifton was started. The town was platted out the next year. It was renamed Shavano, for the mountain on which it was built. The town sprawled over 120 acres, with scattered cabins and a three-story mill across Cyclone Creek on the far end of town. There were a few smelters built, a couple of general stores, and a saloon. All of the buildings were made from logs. The town had been platted out in narrow lots that were only 25 feet wide. The land, wood for building, and water were given away, as long as the new owners agreed to grade the road in front of their houses. Soon Shavano boasted about 100 citizens, and everyone seemed to be optimistic about the future for their town. Then, the mining in the area began to slow down.

The Colorado State Business Directory for 1882 reported the population at 75 with regular stage service by hacks.

By 1883, barely four years after gold was first discovered, the town of Shavano was deserted.

Above the town, by the timberline, is the Billings Tunnel. The face of the tunnel was framed by stones, perfectly cut by a master craftsman.

Unfortunately the mine went dry. In 1904, on New Year's Day there was a silver strike above Shavano, but it was short-lived, and the area is now completely deserted with only scattered ruins.

The legend of Shavano speaks of a young Indian woman whose people had suffered a long, hard drought. The drought forced many of her tribespeople to leave their homes. Every day she prayed at the base of Mount Shavano. Until one day she believed the spirits had spoken to her. By sacrificing herself, the spirits said, she would end the drought for her people. On her death, she became the "Angel of Shavano." There has never been another drought in the Mount Shavano area. She returns every spring and summer, and her tears fall on the land beneath her, making the earth fertile and rich.

STONEWALL
Chaffee County

Directions: Stonewall is slightly less than a mile south of Hancock. When you see the Stonewall Mine, you are there. This town of Stonewall is a different town than the Stonewall in Las Animas County.

Stonewall was formed around the Stonewall Mine about 1879 and merged into the south edge of Hancock. The small town of Stonewall lasted until 1915. It had only one store, one saloon, and depended on Hancock for many of its needs.

After the mine dried up and shut down, the camp disappeared. One source says that it disappeared a decade before Hancock did.

The last building standing at Stonewall, the little town that grew up around the Stonewall Mine about 1879.

TURRET
Chaffee County

Directions: Turret is 12 miles north of Salida. It is a 30-minute drive over well-maintained roads. Coming from the north edge of Salida on Colorado Route 291, you go north on County Road 175, which is also called Ute Trail, for seven miles. At this point you will be at Midway Springs. Turn left on Forest Route 185, and go about a mile and a half. Then take Forest Route 184 on your left, for two miles.

The first activity in the area of Turret was a little distance away down Cat Gulch. The settlers here supported themselves by chopping wood, bringing it down to Nathrop, and turning it into charcoal. In 1881, the Calumet Iron Quarry was opened two miles south of the future site of Turret. A variety of metals and minerals were mined out of the quarry. To serve the quarry, the D&RG Railroad opened a branch of the railroad that went up Railroad Gulch the same year. The railroad, built on too steep a slope, was washed away by a flash flood in 1901. This railroad was never rebuilt.

In 1896, the Gold Bug and the Vivandiere mines were discovered. Miners working here settled into a small camp.

In May 1898, Pete J. Schlosser and the current residents of the area planned the town of Turret. It was named after the mountain that loomed over the town. There is a high natural rock formation on the mountain above. The miners kept looking for the big strike, but it never came. Turret Avenue, the one main street, was crossed by North Russell Street. Some of the key buildings were:

- The Gregory Hotel, a two-story hotel with a balcony, was the most striking building on Turret Avenue. It was on the east side of town, lot 9.

- Turret Hotel and Drug Store was on the southeast corner of Turret Avenue and Main, lot 13.

- The Turret School House was on Turret Avenue on the east side of town, lot 12. It was a two-room schoolhouse, with the teacher's living quarters in the back. After the school was closed, the building became the Bassham Cabin & the Vivandere Barn.

- The Pete Schlosser House was on J. Speer Avenue, lot 4.

- The Pioneer Store and the Gold Belt Newspaper were on Turret Avenue, lot 6.

- Emil Becker's duplex was on South Spring Street, lot 18.

- The Rorabaugh Cabin was on Turret Avenue, the last lot on the eastern edge of town, lot 10.

- The Austin Cabin, built in the late 1890's, was on the corner of South Russell St. and Turret Avenue, lot 16A.

- The Sampling House Saloon site was on Turret Avenue, lot 8. It is no longer standing.

- The Post Office was on the corner of Curtis St. and Turret Avenue on lot 11A.

- The Cemetery was on the southwest corner of the town, lot 8A.

In 1898 the Denver & Rio Grande built a railroad that went to the neighboring town of Calumet five miles to the east. It was different from the railroad in Railroad Gulch two miles to the south. The miners in the area hauled their ore in wagons to the railhead in Calumet, where it would be shipped to Denver.

In 1899 the population peaked at 500. A stage from Salida came into town twice a week. On the Fourth of July in the year 1900, the whole town got together and had a huge celebration. Illinois Senator William Mason spoke to the partying crowd.

At the end of 1900, many of the mines started to dry up, and the residents gradually moved away to more prosperous mining towns. The Independence Mine was the only mine that continued to produce until 1916, but this mine could not keep the town alive.

During the Great Depression, a few poor families lived in the abandoned cabins. They hunted for food and tried to dig up gold for money, but they moved on when the economy recovered.

The Post Office closed its doors for the last time on 12 November 1939. Pete Schlosser, the postmaster, died on 1 May 1939, at the age of 74. The U.S. Post Office closed the Turret Post Office on Pete Schlosser's birthday.

More cabins are scattered around. San Isabel National Forest surrounds the town.

VICKSBURG
Chaffee County

Directions: Take U.S. Highway south from Granite for about three miles until you reach County Road 390. Turn right into Clear Creek Valley. Vicksburg will be about eight miles from U.S. 24 going west. As drive west up the valley, Belford Falls is to the south.

Vicksburg was established in the early 1880's as a mining town. The population peaked at 250, and records show about 200 mining claims were filed in the hillside.

The post office was established in May 1881, but closed only a few years later, in July 1885.

The Colorado State Business Directory for 1882 listed the following on Vicksburg:

VICKSBURG
Mining camp in Chaffee County, 12 miles from Granite. Daily stages. Population, 150. Distances: To Winfield, 5 miles; to Rockvale, 2; to Granite, 12.

Abbott Mining Co.
Glen Cove Mining Co.
Mechanics Mining Co.
Peck, Geo. W., saloon & hotel.
Pelton, J. A. & Co., saloon.
Proteau, Jos., blacksmith.
Rayne, Geo. T. & Co., groceries, hardware, etc.
Reed, Geo. F., gen'l mdse. (Silverdale)
Saddle Mt. Mining Co.
Silverdale M. M. & Imp. Co., Dr. F. H. Wray, supt.
Vicksburg House, Mrs. M. Hoffman.
Washburn, E. A., Assayer.
Woodward, Luke, propr. St. James Hotel.

The town is in an aspen grove next to Clear Creek Gulch. A few of the old cabins have been remodeled and made into summer residences.

Vicksburg had at least 40 buildings. The first big mining interests in the Clear Creek area were started by Governor John Evans, Colorado's second territorial governor, and William Byers, his friend and founder of *The Rocky Mountain News*,. These two men established the La Plata Mining District in 1867 along Clear Creek Gulch, but then they soon lost interest and moved on to greater challenges.

During the winter of 1879, prospectors moved into the area. They found good ore in the creek bed. For 12 miles on each side of Clear Creek Gulch, prospectors worked the ground, and men filed claims as fast as they could. The Ankland, the Baracota, the Iron Cross, the Revenue, and the Swiss Boy were the main mines. There were also some nice farms in the area.

When the mines stopped paying off, the residents left.

WINFIELD
Chaffee County

Directions: Leaving Granite take U.S. Highway 24 for a few miles to County Road 390, then turn right on the County Road. Winfield is exactly 15 miles from Granite. From Buena Vista, Winfield is 25 miles north. Look for a sign on U.S. Highway 24 marking the exit to Winfield and Vicksburg.

Winfield was founded in 1880, after two prospectors chased their runaway burros up the gulch to the furthest northwest corner of the county. The exhausted prospectors were resting at the base of Middle Mountain when they noticed the distinctive glitter of silver. The men staked a few claims, and had the ore assayed, which proved that there was good silver here.

In July 1881, the post office was established, and cabins began to spring up all over. *The Colorado State Business Directory* for 1882 described Winfield:

WINFIELD
Mining camp at the forks of Clear Creek, Chaffee County, 16 miles from Granite and 4 from Vicksburg. Mails tri-weekly. Population 75.

Chicago & London Tunnel Co.
D. & R. G. Mining Co.
Leasure, Thomas, saloon.
Lerian, Adam, general mdse.
Ohio Consolidated Tunnel Co.
Sullivan, Mrs. Orion, hotel and p.m.
Symons, P. H., surveyor.

In the late 1880's the population peaked at 1,500. Many of the men here worked at the Banker Mine. The Concentrator works here did very well, shipping out about a hundred tons of ore a day.

Winfield lasted only a few short years. There is still silver here, but after the silver crash in 1893, no one really considered mining around here. The cabins and the buildings were spaced out so that if a fire raced through town the buildings would be safer. The view of the mountains is beautiful, and Clear Creek is very close. Winfield is below the timberline, located on a mountain meadow, and there are still some buildings remaining. When you are driving on the main road from Beaver City to Winfield, be sure to take the road that appears to be more used.

94 (alternatively, the name is spelled out)
Clear Creek County

Directions: From Idaho Springs go west on U.S. 6 for a couple of miles until you reach County Road 275. This is the Fall River Road, Exit 238. Take this road. After seven miles you will pass the town of Alice, take a four-wheel drive road, on the right. At the first fork stay left, and at the second fork stay right. After this second fork you will have gone about eight miles, and you will be at the site of 94. It will be right side of the trail on the slope of Yankee Hill.

St. Mary's Glacier is a few miles northwest of the town, and the town offers spectacular views of mountain peaks reaching 12,000 and 13,000 feet.

On the western side of Yankee Hill, almost on a ledge built in the hill, the town of 94 came to life. The miners named their camp after the 94 mine, which had been named for the year it was discovered, 1894. The town is one mile past Alice and one mile before Yankee. Although it was never officially incorporated, and it did not have its own post office, 94 grew into a town in its own right. Cabins, houses and other buildings were scattered on the hillside. Townspeople worked at the 94, the Lalla, and the Princess Alice mines. Ninety-four relied on the town of Alice, a mile to the southwest, for supplies.

The little town was difficult to reach in the best of weather, and in winter it was almost totally isolated by blizzards and blowing, drifting snows. When the mines stopped producing, the residents left, and little remains of the town.

ALICE (Yorktown)
Clear Creek County

Directions: Alice is ten miles from Idaho Springs. From Idaho Springs, take I-70 west for a little over a mile until you reach the Fall River Exit, exit 238. Take this exit. This is County Road 275. Take this for about nine miles to the town of Alice. This is a good road, which nearly any car can make.

Just north west of Idaho Springs and next to the Fall River is the town of Alice, originally called Yorktown. A local prospector named Taylor discovered a rich gold mine, which he named Alice, after his wife. Before long, the little town was called Alice after the mine.

A reporter for Denver's *Republican* newspaper interviewed one of the colorful old miners who was connected with Alice:

Surrounded by a half score of old mining cronies, and indulging in reminiscences of old mining days, a Republican reporter found the hon. A. W. Callen of Plaruias, Arizona, last night. Mr. Callen is better known as "Old Grizzly," the cognomen being attached to his personality because of the grizzly grey beard which surrounds his face, standing out in hazel-brush style and white with age. "Old Grizzly" was born in New York 70 years ago, and wears his age lightly for a man who has passed through so many vicissitudes. He has been lawyer, legislator, merchant, miner and a little of everything else. Has been up and down as almost every man has who has followed mining. "Old Grizzly's" remarkable career as briefly sketched to THE REPUBLICAN reporter last night has been a very interesting one. He was educated at Centerville, New York and graduated at Coalbough's law school.

A Long Time Ago.
He came West at the age of 22 and was for several years in the employ of the Green Bay Fur Company, trading among the Winnebagoes and the Sioux. He was married at the age of 28 to an old school-mate and has raised a family, consisting of two sons and two daughters, all of them living in Junction City, Kansas, except one son, the noted Jim S. Callen of San Diego, California, who

stumped the State for Harrison and lost a Democratic partner as a result of his hard work for the Republican ticket.

"Old Grizzly" was a Whig until the organization of the Republican party, when he joined the Republicans, and has voted with them ever since. He was a member of the Kansas legislature for three terms, has been sheriff, probate judge, member of school boards, and held about all the local offices provided for by the Kansas statute. He was a '49er, was one of the first in at the Pike's Peak excitement, and joined the rush to Leadville. In 1873 he took an expedition to Arizona with his own teams and wagons and men hired for one year, but the noble Red Man took a hand and the expedition proved a failure. He returned to Colorado in 1875 and located some mines, which he still owns. Among these are the Little Prince and the Ada and Alice. He operated in Leadville in 1881 and 1882 and was a member of the firm of Callen, Strickler & Company.

The Cost of a Trip.

Along about that time he took a trip to New York, and dropped about $300,000. He returned to Arizona in 1883, and has remained there since that time with the ups and downs incident to a mining life.

"At the present time," said Mr. Callen, "I am down, and owe every man in Arizona and a good portion of the population of Kansas, but am still in hopes of making a raise and being buried in a fine metallic casket at Junction City, Kansas." Mr. Callen now owns the Placerita and Carbonate mines, thirty five files from Prescott, Arizona, and numerous smaller mines. He is in Denver now with J.W. O'Bryan, a noted mining man, who visited the Arizona camps in the interest of himself, C.W. Badgley and other mining men of Colorado. Mr. O'Bryan brought back with him half a ton of samples from about twenty mines, all of which have been assayed and run from $20 to $6,700 to the ton in free gold, which naturally makes "Old Grizzly" feel good.

The Denver Republican
Tuesday, 8 April 1890

Alice was the largest mining camp in the Fall River area. Nearby were the towns of Fall River and Silver City. Tents and cabins lined much of Fall River. The Alice Mine had its own mill, and just past the town was the Alice Glory Hole, a vast pit about 150 feet wide and almost 50 feet deep, producing mainly gold with a little silver and lead. The rich placer gold was mined with hydraulic methods, which stripped away much of the surface of the surrounding area.

When the Alice Mine closed down in 1899, nearly everyone in the town packed up and left within a week. Townspeople abandoned Alice so fast that the town officials forgot to give their schoolteacher her last paycheck.

Today a few old cabins and the little white schoolhouse still stand, and there are a few scattered houses and summer cabins.

The original Alice schoolhouse, now boarded up and empty. When the Alice Mine closed down in 1899, the townspeople left in such a hurry that town officials forgot to give their schoolteacher her last paycheck.

BONITO
Clear Creek County

Directions: Take the four-wheel drive road west from the Freeland site for about a mile. Freeland looks like nothing so much as a cluster of tailings piles. Today, it is difficult to find any trace of Bonito, but if you look very closely, you might discover an old cabin foundation or two.

Bonito was a suburb of Freeland that grew up as a home for the area miners. Bonito also had a few mills that processed ore from the area mines.

In 1884, the Bullion Smelter opened in Freeland, employing 25 men.

When hiking in the area in the summer of 2000, the authors found it hard to believe that a community once thrived here.

FREELAND (Trail Creek Camp)
Clear Creek County

Directions: You should have a four-wheel drive vehicle for this trip, although the road is not that bad in the summer. From Idaho Springs, go west to the well-marked Trail Creek Road. After a couple of hundred feet you will pass the Phoenix Mine. Continue on Trail Creek Road for about four miles. Although Trail Creek is a ledge road, the area is so heavily forested that it is manageable for most drivers and queasy passengers. After about four miles, you will come upon the tailings piles that were once Freeland.

In the early summer of 1877, after a few rich silver discoveries the little town of Freeland got its start. Freeland was on the Fall River about three miles from Idaho Springs. The Clear Creek Stagecoach brought visitors and supplies. In 1882 according to the *Colorado Shippers Guide*, it had a population of 450. The town was named after the main mine in the area, the Freeland Mine. The Freeland Mine was discovered by John Dumont, the man for whom the town of Dumont in Clear Creek County was named. A couple of years after discovering the Freeland Mine, Dumont sold it to John W. Mackay, a former owner of the legendary Nevada Comstock Mine. Mackay invested his Comstock earnings in the Freeland. Lone Tree Mine was second to the Freeland in production.

Many local miners and prospectors gave up their independent ventures for the security of a paycheck working for Mackay. The little town of Bonito grew up next to Freeland, and its residents worked in Freeland's mills and mines.

Freeland had its own public school for its children. At least 80 homes had been built.

In 1882, *The Colorado Business Directory* reported the following;

FREELAND
Mining Camp in Clear Creek County, 3 miles south of Idaho Springs. Daily Mails.

Armstrong, J.D. & Co., gen'l mdse
Baker, C.A., meat market
Banner Group Mining Co.

Coze, Geo., saloon
Freeland Consolidated Mill
Lone Tree Mining Co.
Moss Mining Co.
Stratton Mining Co.
Trevillion, Wm., saloon

In 1884, the Freeland Mining Company acquired the Pacific Mining and Reduction Works in neighboring Idaho Springs.

The Clear Creek and Gilpin County Directory for 1892-3 was published by the Bensel Directory Company. It named the following residents of Freeland:

Anderson, Charles, lab
Anderson, Fred, miner
Armstrong, Lon, miner
Bachelder, William, miner
Berg, Charles, miner
Bone, John, sr., miner
Bone, John, jr., miner
Bunt, John, miner
Claus, Fred, miner
Cox, John E., miner
Davey, John A., miner
Defrance, Fred, miner
Dodge, John, teamster
Doyle, W. A., miner
Gilbert, J. G., miner
Goyne, I. D., miner
Harper, Lewis, teamster
Hawke, William, miner
Head, John, miner
Johnson, A. J., miner
Johnson, Victor, miner
Joseph, Howell, lab
Joseph, John D., miner
Libby, Richard, miner
Lumley, David, miner

Miller, Charles L., miner
Moore, H. N., miner
Morgan, John E., miner
McKenzie, Duncan, chopper
McKenzie, John A., chopper
McKenzie, R. W., teamster
Nicholls, W. H., millman
Parlin, David, miner
Pohnear, John, butcher
Roberts, Jenkin, postmaster and dealer in general merchandise
Snell, James, miner
Stephens, John, saloonkeeper
Stephens, William, miner
Strom, John, miner
Tonkin, Samuel, miner
Traylor, Eddy, miner
Traylor, William, miner
Twistelton, William, miner
Williams, E. C., miner
Williams, Richard, miner
Williams, Thomas, miner
Williams, William, miner
Williams, W. H., capitalist

The Colorado Business Directory for 1896 listed the following for Freeland:

FREELAND
Mining camp in Clear Creek county, 5 miles west of Idaho Springs. Daily mails. Population 400. Altitude 9,000.

Bone, John, boarding
Johnson, Mrs. A J, boarding
Joseph, John, boarding
Libby, Mrs. E, boarding
Palmer, John, meat market
Robert, J, gen mdse and postmaster
Stephens, John, saloon

In the early 1900's the area mines started to play out, and, one by one, they shut down.

By the summer of 2000, the authors were unable to find any existing buildings from this once bustling community. If you continue up this main road, it will take you to the Lamartine Mine.

LAMARTINE
Clear Creek County

Directions: You should attempt this trip only in the late summer and only with a four-wheel drive. Lamartine is very hard to find, and there are three different ways to get there.

1. The Spring Gulch Route. Drive through Idaho Springs heading west on the main road. Go straight until you reach 13th Street. At 13th Street, turn left taking a bridge over a highway that turns into 103. Then to the right you turn on to Spring Gulch Road, which is a little bit before the Idaho Springs Cemetery. Spring Gulch Road is a dirt road with a sign saying that this is a residential area for the next five miles and that there is no trespassing on private property. There is a 15- or 20-foot dropoff on the left. A few houses are scattered on the left side of the road. On the right side of the road is a mountain, with a couple of mine openings. We noticed a very old narrow cabin pressed against the mountain and almost touching the road, and on the left side of the road is an old building that belongs to the old long cabin. Keep driving up Spring Gulch, through the forest of aspens and pines. The road will start to get rough. About four miles up, the authors parked and hiked the rest of the way. At this point, the road branched out in five directions. We took the main trail to the right that went upwards. The road is grooved and rutted from harsh weather and vehicles. We walked this road for a while, and then there was a steep road coming up from below on the right. A sign at the crossroads pointed to the road where we came from saying Spring Gulch, and then another sign said Trail Creek, which was on the road that had just joined us. Continue on the path, there will be many trails and roads leading back and forth. Lamartine is over the ridge of the mountain in a high alpine meadow. Stay on the most traveled trail. You will notice power lines following the trail for a while. We kept walking, soon to the left the mountain ridge was only 100 feet up, and to the right was a city of tailings, the remnants of Freeland. There were no discernable remains of a building. Continue on the main trail, which has now crossed over the ridge and is starting to descend the side of the mountain and extend to a clearing. It took us less than ten minutes to reach the lower part of the road. A four-wheel drive dirt road to the left joined the road that we were on, and a sign identified it as the Ute Creek Road. We continued walking until the road forked; stay right. The road is very steep. After about a hundred feet we saw our first log cabin foundation to the left next to a small creek and a tree line. Broken pottery and shards of purple glass were strewn all over the ground if you looked closely. This was most definitely the sloping meadow, and many more foundations were found among the trees. In the underbrush, we noticed rocks outlining the remains of long-ago homes and cabins. In this little area, we were able to identify eight foundations and saw traces of many more in the trees.

2. The Trail Creek Route. Go to the western edge of Idaho Springs and turn on Trail Creek Road. Stay on Trail Creek Road for about three miles. You will pass the site of Freeland, marked by its many tailings piles. Stay right when the road forks, and stay on the most traveled road. The trail gets steeper and rougher. A century ago, this was a toll road. You will soon pass a little cinder block building, possibly the old tollgate building. An old metal pole in front of the building is evidence of an old gate. Not far from here on the road is the Lamartine Mine, with its mine and mill buildings. The stamp machinery was still in place as recently as August 2000. The road continues, and is steep, rocky, and rutted. Follow it, and take the third trail on the right, to the alpine meadow that is the town site for Lamartine.

3. Third Possible Route to Lamartine. Take 103 out of Idaho Springs for about five miles until you reach Ute Creek Road. Take this west for a mile and a half. Then you take a four-wheel drive trail for about a mile or so until you reach the site. It spans a meadow and stretches up the side of the mountain.

In the early 1870's four men from Idaho Springs, Chavanne, Cooper, Bougher, and Medill, discovered a little rich gold float in Trail Creek. The men followed the creek upstream to the top of the mountain, to discover the source of the float. There, they discovered a large gold vein. They staked their claim together, naming it the Lamartine. They did very little to develop their property. Not long after the claim was staked, Bougher died. His wife was so desperately poor that she sold her husband's share of the claim to Peter Himrod, her brother-in-law, for $250. Cooper, also in need of money, sold his portion of the claim to Peter Himrod for $25. The Chaffee Act passed by Congress in 1872 required mine owners to put at least $100 worth of work each year in the claim or they would forfeit their claim. When the third partner, Medill, left the area, abandoning his portion of the claim, he legally forfeited it. That left only Peter Himrod and Chavanne as joint partners of the Lamartine.

One day Peter Himrod went to visit the claim, taking Trail Creek Road, then a toll road. At the time Chavanne worked as the tollgate keeper, and he was also in need of money. When Chavanne saw Peter Himrod coming up the toll road, he cornered his partner, and asked him to buy his portion of the Lamartine. Himrod did not want the other half of the claim, but he paid Chavanne $5 for it so that Chavanne would leave him alone. Himrod now owned the entire claim for $280. He did not think the mine had any real value, did not invest anything to develop it, and soon returned to his home in New York.

Since he lived in New York and had little interest in working the Lamartine claim, Peter Himrod agreed to lease it to Frank Craig and W. S. Comier.

Peter Himrod died soon after this agreement with Comier, leaving the claim to his son, who may have been named Fred. Although rich silver and gold ore had been found, his son did not want to work it and sold the claim for $360. Almost immediately, the new owners developed the mine, and by the turn of the century it produced at least $3 million in gold, silver, and lead.

The town of Lamartine, named after the Lamartine mine, grew up and soon boasted a peak population of 500. It was built on a high alpine meadow with half of the buildings built on the side of the mountain like stair steps. There was a public school, a store, and a saloon. On the mountain across from Lamartine, was a small town, now just a site, of woodcutters with their own mills who supplied wood for the smelters, cabins, and buildings.

After the turn of the century the Lamartine mine closed down, and the residents moved away. Today, Lamartine is a forgotten town consisting of cabin foundations. In the neighboring town of Idaho Springs, few people know of Lamartine.

The Bensel Directory Company published the *Clear Creek and Gilpin County Directory* for the years 1892-3. It named the following residents for Lamartine (interestingly, no women are included on the list);

Anderson, J. R., miner
Brown, Leander, miner
Brown, Richard, miner
Brubaker, H. E., engineer
Burns, Fred C., miner

Catlin, L. C., teamster
Cole, George A., engineer
Cook, J. E., miner
Cox, George A., engineer
Davis, John C., miner
Dowe, Archei, miner
Feeley, Martin, miner
Fitzmorse, Thos., miner
Gable, William, miner
Gates, Theodore, fireman
Greenhow, Thomas, teamster
Gunstrom, John, miner
Harper, Stephen, miner
Harrington, Dennis, blacksmith
Hicks, J. M., miner
Irish, D. C., merchant
Johnson, Alfred, miner
Kearns, John, miner
Lang, John, miner
Lanning, O. E., miner
Lundberg, A., miner
Lundk, G. G., carp [carpenter]
Lundk, P. G., carp [carpenter]
Marshall, Graham, miner
Marshall, G. G., merchant
Marshall, James Y., miner
Marrill, Wiley, miner
Mosher, W. H., teamster
McFarland, D. L., machinist
McGuire, John, miner
O`shea, T. E., miner
Phillips, John A., miner
Phillips, William M., miner
Potts, R. F., teamster
Rowse, Charles, miner
Roy, J. B., miner
Strom, Charles, miner
Thomas, C. J., miner
Thomas, G. C., miner

Thomas, T. S., miner
Tremboth, Harry, miner
Von Brandis, C., blacksmith

The Denver Republican printed the following article about Lamartine, Saturday Morning, 18 January 1890:

UTE CREEK COUNTRY.
A Coming Camp -The Famous Lamartine Mine. Etc.

The Ute creek district is six miles from Idaho Springs, and only 44 miles from Denver. There is no other portion of the State where fissure veins abound which has a more promising future than this district. It is here that the famous Lamartine mine is situated, one of the best paying, if not the best true fissure vein in the state. A vast tract of territory surrounds this mine, which with rare exceptions, has yet to receive a mere scratching from the prospector. Ore lies as a general rule, at a depth of 100 to 300 feet, and when it is struck it is rich. All ore is shipped to Idaho Springs over a wagon road, open all the year. The Lamartine is an old patented lode and remained idle for years. In the spring of 1887, Frank Craig and W. S. Comier, whilst prospecting found some rich float and traced it up to where the present No. 2 shaft is sunk. They made a location and started to sink, when they were informed that they were on patented ground. The best prospectors will make these mistakes. The owner of the Lamartine, Peter Himrod of New York City, then gave them contracts to sink the shaft. No ore was struck until a depth of 125 feet was reached, when a small pocket was encountered. Sinking was continued and at 250 feet the big pay was struck that held out for nearly two years, and at the present time there is more ore in sight than ever before. Twelve ore sorters are at work, making two classes of ore. The mine is worked through three shafts, the deepest one is 365 feet. No. 3 shaft has one of the finest plants of machinery in the country. This shaft is a double compartment one and is sunk with steam drills. A cage is in use in this shaft. The output of the Lamartine is about $75,000 a month. Over $1,000,000 has been taken out in eighteen months. The present lessees, Burns, Hatchet, Armstrong, and Williams, employ 125 men, who work 8 hour shifts at $3 per day, sublessees, who pay 50 per cent royalty, cleared for the month of December $1,500 per man. Ore is being turned out faster than the teams can haul it away. From 12 to 16, 4 horse teams are kept busy every day of the week. The Lamartine, like all rich mines has a thousand and one extensions. On the east is the El Dorado, the shaft which is being sunk as fast as two shifts can do it. It is now down about 270 feet with scattered ore at the bottom. The Ben Harrison Shaft is also being sunk. An engine is to be placed over this shaft in a few weeks. For a prospect vein shows up remarkably well. On the west is the Flancier, which is supposed to be the genuine extension of the Lamartine. The Bell and Redd shaft is now down about 250 feet, with three shafts still sinking, quartz and ore mixed in the bottom. This mine produced some very rich surface quartz. The ore shaft on this vein is being sunk within a few feet of the Lamartine end lines. The Lamartine "pocket" is expected to be struck at about 150 feet. A little north of the Flancier is the Monnie Musk, with a shaft down nearly 200 feet and still sinking. Some rich ore is exposed in the bottom. A fine plant of machinery will be erected next month on this property. The Chloride Hill, a fine looking property north of the Lamartine, has been leased and bonded to St. Louis parties, who will sink the shaft 100 feet deeper. Up the hill from the Monnie Musk is the St. Louis parties shaft down 225 feet, and still sinking. Good ore was recently struck in the shaft. This is a good property.

The Lost Treasure, a prospect 40 feet deep, was recently sold to Denver parties, the consideration being $3,000. A contract has been let to sink the shaft 100 feet deeper. The Albany, Hawkeye, Denver City, Union and many others lying in the vicinity of the Lamartine are all good properties.

Mining investors would do well to visit the Ute creek country and examine it thoroughly.- It is going to be a great camp. There are a few more Lamartines up Ute if capital will only take hold and help.

The Denver Republican
Morning, 18 January 1890

NORTH EMPIRE (Upper Empire)
Clear Creek County

Directions: Empire is on 1-70. Coming from the town of Empire, take a four-wheel drive dirt road directly north of town for a little over a mile, and this will take you directly into North Empire. North Empire was slightly more than a mile above the town of Empire.

North Empire came to life in 1860 when Clinton Cowles, Henry Dewitt, and Edgar Freeman discovered wire gold. Many good mines were soon discovered. The Negus Placer Mining Company produced $50,000 in gold in 1878. That same year hydraulic mining had begun in North Empire. The Conqueror Mining and Milling Company was the biggest and the most productive. North Empire, basically a suburb of Empire, became a boomtown in the 1860's. Its population was constantly changing until about 1940 when the town was abandoned. The town consisted of a few saloons, some boardinghouses, a general store, a couple of mills, and many cabins.

The Caribou Post printed the following article:

MINING AT EMPIRE
A visitor to the mines in that district reports to us that the Callery Brothers have reached a good paying vein on the Cashier lode, on Covode Mountains, near Lower Empire. An adit has been driven in on the vein sixty feet, showing in the face eight inches of first class ore that averages five ounces of gold and thirteen ounces of silver per ton, and twelve inches of second class ore. The lode between the walls is six feet wide. The ore is sold to the Stewart Reducing Company. Four miners are employed in the mine. Thus will Empire again come out of her embarrassments little by little until finally the district will again contain her thousands of miners and mill men. In a few days we hope to be able to give our readers some items on the Crosby Furnace Workings.

The Caribou Post
Saturday, 5 August 1871

Recent mining activity in the area has been minimal, not enough to reestablish the town. North Empire is known for a toboggan run that went from North Empire to Empire a little more than a mile below.

RED ELEPHANT
Clear Creek County

Directions: Red Elephant is one mile north of Lawson on an unmarked four-wheel drive road. It is on Red Elephant Mountain, 47 miles from Denver.

In 1865, a prospector on Red Elephant Mountain discovered some loose blossom rock, which he had assayed at $18,000 a ton. Prospectors never found the rich source of this blossom rock. Many other good veins were discovered, the best assaying at $2,500 a ton, but none were nearly as rich as the blossom rock. A little town soon grew on Red Elephant Mountain, called Red Elephant. The little town became a company town operated by the Red Elephant Mining Company.

In 1882 *The Denver Business Directory* reported that Red Elephant had a population of 200:

RED ELEPHANT
Post-office in Clear Creek County, Georgetown Branch Colorado Central Railroad.

Red Elephant Mining Co.
Wolfe & West, groceries.

Three years later, *Crofutt's Grip-Sack Guide of Colorado* reported that the town had grown to 300 people, with a post office, general store, many boardinghouses, and cabins. One mile south of Red Elephant is the town of Lawson. Red Elephant was basically a suburb of Lawson. Lawson was the closest train stop, so transportation was never a problem.

Eventually the town of Red Elephant was abandoned, and some say that the population probably moved to nearby Lawson. Not much is left of Red Elephant.

SILVER CREEK (Daileyville, Chinn City)
Clear Creek County

Directions: The little town of Lawson is on I-70. From Lawson turn south on a four-wheel drive dirt road that goes uphill for two miles. Silver Creek is about ten miles northeast of Georgetown.

On a southern mountaintop in Clear Creek Canyon, the little unincorporated town of Silver Creek sprang up in the early 1880's. There was a flurry of mining here before 1880, but the residents did not consider their community a town until 1884. The town consisted of a school, a post office, a saloon, a general store, many cabins, the Reynolds' Mining Properties, and the Seven Sisters' Mining Properties. The area mines produced silver and lead. In the beginning the community was nicknamed Daileyville after a local man. Some sources say that at the very beginning of Silver Creek's life that it was also called Chinn City

The Bensel Directory Companies published the *Clear Creek and Gilpin County Directory* for the years 1892-3. It named the following residents of Silver Creek:

Anderson, Eli, miner
Billings, C. L., miner
Chrisholm, Malcolm, teamster
Clancy, James, miner
Crosby, J. E., miner
Dailey, John, miner
Daily, Patrick, miner
Doyle, James, miner
Duloney, David, miner
Duncoff, Thomas, miner
Friddich, Jacob, miner
Gates, John S., miner
Harding, George, miner
Hurley, William, miner
Johnson, A. C., miner
Lucha, William, miner
Morrison, Alex, miner
McDonough, John, miner
McLaughlin, H., miner
McMillian, George, miner

Nelson, A. E., miner
Ollie--, miner
Ollie--, miner
Powell, George, miner
Rodda, Thomas, miner
Rodda, William, miner
Sebastrom, Pratt, miner
Tranesch, George, miner
Whalen, Maurice, miner
White, Patrick, miner

Silver Creek had a brief life. By the early 1900's, the mines played out, and the town was abandoned. If you can get up to Silver Creek by four-wheel drive, there is a trail that continues south that will take you to Lamartine.

There is almost nothing left.

SPANISH BAR
Clear Creek County

Directions: Spanish Bar was two miles west of the Idaho Springs city limits on Clear Creek, just before the Trail Creek Road, right at the junction of South Clear Creek and Fall River. The 1884 *Croffut's Grip-Sack Guide of Colorado* reports that Spanish Bar was only 39 miles west of Denver.

Prospectors in the 1860's discovered rich gold float along Clear Creek, two miles west of Idaho Springs, then called Jackson's Diggings. Prospectors lined south Clear Creek panning gold. Spanish Bar was an unincorporated little community. The origin of the town's name is unclear because no Spanish surnames appear in early Federal Census records for Spanish Bar. The little mining town may have been named Spanish Bar for the early Spanish explorers and their bars of Spanish gold.

At its peak in 1880 with a population of 400, Spanish Bar had many cabins, houses, businesses, mills, and mines. By 1882, the *Denver Business Directory's Shipping Guide* reported that the population had fallen to only 100. The town had regular stagecoach transportation to Georgetown. Three years later, in 1885, *Crofutt's Grip-Sack Guide of Colorado* reported that only 50 people lived in Spanish Bar. The town was only half a mile from the Freeland Mill.

By 1890, the town was deserted.

The site is now a grassy meadow along the south edge of I-70 by the Stanley Mine.

In 1880, about 400 people lived in Spanish Bar. Today it is only a grassy meadow and some tailing piles. In this view of the site, Stanley Mine is in the center of the photo, and you would never guess that a little community once thrived here.

WALDORF
Clear Creek County

Directions: Waldorf is about five or six miles south of Georgetown on Forest Route 248, via County Road 381, passable by any car. It has an elevation of 11,666 feet.

In 1868, silver was discovered at the eastern base of Argentine Pass. A very rich vein of silver sulfide was dubbed the Stevens' Lode. Old tales describe how mineworkers wrapped the ore in cowhide bags and rolled it down the mountainside to transport it. They did it this way until they realized that they were losing most of it. Just a few years after the Stevens' discovery, 75 claims had been staked out in the area. Only nine of these claims were successful. Before you knew it, a little mining community named Waldorf sprouted on the side of Mount McClellan. Edward John Wilcox owned most of the mining properties in the area and formed the Waldorf Mining Company. By consolidating his holdings, he transformed Waldorf into a prosperous company town, an Edward John Wilcox company town. Wilcox was a good hardworking man, who had once been a Methodist Minister. Wilcox took care of his employees, his town, and then on August 1, 1906, he had finished building his railroad, the highest multiuse steam engine railroad in the world. It was called the Argentine Central Railroad. With full 145 degree turns, it went from Silver Plume to Waldorf, then another two miles above Waldorf to the Santiago Mine, and from there to the top of Mount McClellan where there is a frozen cave featuring glittering icicles. Here, the train made a brief stop so the passengers could get out, see the 106 visible mountain peaks, play in the snow, and visit the ice cave. This became such a tourist attraction that Wilcox started to charge admission.

Waldorf was a busy little town, boasting a boardinghouse, hotel, stables, a mill, a post office, a powerhouse, a machine shop, and a few other businesses. Cabins were scattered throughout the area.

Wilcox's mines paid out in excess of $4 million before they stopped producing, causing the town to become abandoned. The town has since fallen down upon itself, but there is a sign marking the site of Waldorf.

YANKEE   (Yankee Hill)
Clear Creek County

Directions: From Idaho Springs go west on U.S. 6 for a couple of miles until you reach County Road 275. This is the Fall River Road, Exit 238. After seven miles on this road, you will pass the town of Alice. Here, take the four-wheel drive trail on the right side of the road. At the first fork stay left, at the second fork stay right. At this point, to the right is the town site of 94, and you will have gone about eight miles. Continue on the road for another mile, and you will be at Yankee on top of Yankee Hill.

In the early 1890's gold was discovered on Yankee Hill, and a little gold camp was formed. The men called the camp Yankee. The town was on the popular stagecoach route that went from Georgetown to Central City. Yankee soon became a stagecoach stop. Cabins and mining claims were scattered all over the top of Yankee Hill. By 1893, a post office was established, and then a mill was built in town. The best mine in the town was the Gold Anchor Mine. *The Colorado State Business Directory* for 1896 listed the following on Yankee:

YANKEE
Gold camp in Clear Creek county, 6 miles from Central City, 12 miles from Idaho Springs. Pop. 500.

Cheney, Edw F, journalist
Crompton, W G, physician
Dentress, C, hotel
Elliot, John, justice peace
Gow, Wm, mining eng & sur
Hawk, Mrs Loretta G, postmistress
Hawks, A M, supt American Watt M Co
Lake, M C, constable
Lamb, J W, gen mdse
Miller, Arnold, hotel

The town was prosperous until after the turn of the century when its fortunes changed. The quality of its ores declined, and the mines were no longer productive. The town residents started to move away, and by 1910 the post office closed. Almost nothing remains of this bustling little town, and you must look very closely to find traces of a foundation.

RUSSELL (Sangre de Cristo, Placer)
Costilla County

Directions: Russell is 87 miles southwest of Pueblo on U.S. Highway 160, 12 miles northeast of Fort Garland on U.S. Highway 160, and seven miles southwest of LeVeta Pass. U.S. Highway 160 crosses Sangre de Cristo Creek about one mile north of Russell.

The area features a grey modera shale that is rich in marine fossils of brachiopods, corals, gastropods, and small radiolaria.

Beginning about 1852, soldiers from nearby Fort Massachusetts (now, Fort Garland) spent their off-duty hours placer mining at Grayback Gulch, nicknamed Officers' Bar, Placer Creek, and Spanish Gulch. The gulch is just north of Russell.

A mining camp did not develop until about 30 years later during the gold rush. The residents of the camp originally named it Sangre de Cristo, Spanish for "the blood of Christ." Eventually the name changed to Russell.

Hopeful prospectors staked many claims. Silver, gold, and copper were mined, and the town grew quickly. The Colorado Coal and Iron Company owned and worked a few of these claims.

The Colorado Business Directory for 1882, listed the town name as Placer, with a post office called Russell. It reported a population of 550.

PLACER

Station on Denver and Rio Grand Railway, 212 miles southwest of Denver, in Costilla County. Daily mails. Post office name is Russell.

Bohn, John, saloon.
Caldwell, E.K., lumber mill.
Evans, D.T., constable.
Eyssell, Hugo, drugs, stationery, etc.
Farmer, Joseph, meat market.
Graham, Tim, restaurant.
Grant, H. H., saloon.
Hamilton, Ayres, meat market.

Hamilton, A., county judge.
Hampton, A. K., assessor.
Hampton Bros., blacksmiths.
Holt, E. T., dentist.
Marra, John O., boot and shoemaker.
Morton House, G. W. Morton propr.
Morton, G. W., gen'l mdse.
Sefton, H. T., gen'l mdse. and postmaster.
Ward, Peter, restaurant.
Weinberger, L., tobacco and cigars.
Wilcox & Lake, lodging house.
Wolfe & Sutton, millers.

The Denver & Rio Grande Railroad ran a narrow gauge railroad through the town and over LeVeta Pass. When the railroad converted to a standard gauge, it ripped up the tracks and moved the depot a few miles south of Russell. This move isolated the camp, and the mines were not productive enough to save it.

By 1885, the population had dwindled to 200, according to *Crofutt's Grip-Sack Guide of Colorado*. Two years later, only 65 people were left in Russell.

The last residents here tried to save the town by turning to ranching. Their efforts failed because the area was so isolated, and soon everyone left.

ILSE
Custer County

Directions: Ilse is almost 30 miles southwest of Florence and three miles southeast of the site of Galena. From Silver Cliff, take State Highway 96 east for about nine miles, until you reach County Road 271, formerly an old stagecoach road between Canon City and Rosita. Take this for about four miles. Little remains of the town site, making it very difficult to find.

In 1878, a Dutch miner named DeLamar made a good silver strike in the area, which drew other miners to the area to prospect for a rich strike of their own. Within a short time, a second strike was made. This mine, called the Terrible, was the main mine in town employing at least a hundred miners. It was named the Terrible because one of the first miners that took a look at it saw a huge deposit of lead in the mine. Making a face, he exclaimed, "That's a terrible big deposit!" So then the terrible name stuck.

DeLamar discovered the first silver ore in the area, but by the end of the year his mine was exhausted. He was certain there was more silver, he was positive that the mother lode would turn up if he just kept digging. Spending all his money, he started borrowing money from people in town. Still, DeLamar found nothing. Skipping out of town after dark, leaving everyone behind without paying his debts, he went to a little town in Idaho. He still dreamed of finding that mother lode of silver. In Idaho, he continued to prospect and was finally rewarded for all his hard work. DeLamar returned to Ilse and repaid everyone five times what he owed.

The town of Ilse grew to 300 people. Lining the main street were three saloons, a boarding house, a hotel, post office, and a general store.

In 1887, a fire swept through and destroyed most of the town. The mine had not been producing well, so no one bothered to rebuild the town. They simply moved on.

In 1903, a mining company came to the area and built a mill. The company did a little mining, but the town of Isle is gone.

QUERIDA (Bassick City, Bassickville)
Custer County

Directions: Coming from Silver Cliff take State Highway 96 east for six miles, then take County Road 341 south for a mile. Querida will be on the right.

Querida got its start in early January 1877. Edmund C. Bassick was a miner working for the Centennial Mining Company, which operated a productive mine on Tyndal Mountain in what is now Querida. Bassick lived in Rosita and walked to work everyday to Querida. One day, he stopped to shake some dirt out of his shoe, lost his balance, and stepped on an anthill. Within seconds, hundreds of biting ants swarming up his legs. He furiously brushed at the ants, and, still angry, took his miner's pick to the anthill. His pick kept hitting rock, but he kept slashing until he destroyed the anthill. He picked up a piece of the rock, put it in his pocket, and walked home. Back home in Rosita, Bassick finally took a close look at the rock. Could it really be silver? He showed his rock to an assayer, but he could not afford the assayer's fee. To pay for the assay, Bassick cut a cord of wood in exchange. The rock assayed at 109 troy ounces of silver per ton. The silver was very pure, and the rock also showed traces of gold. On 4 January 1877, Bassick staked his claim, calling it the Bassick. Bassick then returned to Centennial Mining Company and quit his job. Bassick sold his first wagonload of ore for $10,000.

The little town that grew up around the Bassick Mine was called Bassickville. As the town continued to grow, it became Bassick City.

Over the next two years, Bassick earned about $500,000 from his mine. Then, in 1879 Bassick sold the mine to an Eastern company for $500,000 plus a ten percent interest in the company. With that, he happily returned to his hometown, Bridgeport, Connecticut, where he bought the house that had once belonged to P.T. Barnum.

After Bassick left, the name of the town was changed to Querida, Spanish for "my darling" or "beloved." One source says that the new owners of the mine renamed the town. Another source reports that the town was renamed by David Livingston, nephew of the famous Dr. Livingston and one of the original prospectors who helped build the town. David Livingston was a Scotsman who migrated to Rosita and then to Bassickville in 1873 and 1874. He returned to Scotland to bring his wife to his new home in 1878.

According to the *Colorado State Shippers Guide*, the town had 150 citizens by 1882. Querida grew into a small city, with stores, a sawmill, a smelting works, and even a three-story hotel. The new owners of the mine focused only on profits. They scripted one publicity stunt after another to get attention, hoping to create demand and inflate the price of their mine stock. One time they spread rumors that the mine was about to be robbed, and then exploded a load of dynamite in the yard in front of the mine office, trying to whip up a commotion. The company owners then decided to sink a huge shaft far from the original shaft. They had hoped to retrieve the ore easier and cheaper by taking this shaft straight down over the vein. Their efforts were unsuccessful, and it ended up costing a great deal of money. This was the last straw for the stockholders, they had had it, and they started court proceedings. Then there were arguments over conflicting claims. In 1884, these problems caused the courts to shut down the mine, which was still very rich in silver. There was reportedly over a million dollars of silver still to be mined.

The following article was printed about the Bassick Mine:

The Bassick mine bids fair to prove equal to the Comstock in its palmist days. The main shaft in the Bassick is now about 1,170 feet deep, and the mine is producing all the ore that the mill can treat, besides a good deal of first-class ore that runs up in the thousands of dollars per ton. The running expenses of the company are from $16,000 to $22,000 a month and the mine has paid $300,000 in dividends this year, and the gross product of the mine for the year is $700,000 or $800,000. – *Pueblo Chieftain.*

The White Pine Cone
11 January 1884

The Bassick Mine was the sole support of Querida and Rosita. Nearly everyone in Querida left. In 1903 the Melrose Gold Mining Company worked the area on and off, but the town is gone. The few families that tried to stay on were all gone by 1906 when the post office was discontinued. There are some building remains left.

The community was destroyed by greed.

ROSITA (Brown's Spring)
Custer County

Directions: Coming from Colorado Springs, go south on State Highway 115 to Florence. From Florence go south on State Highway 67 to Wetmore. Just past Wetmore, take State Highway 96 west until you see a sign to Rosita, On the left, take County Road 341, then take County Road 329 going south. When you reach County Road 328 go left (east) for a tenth of a mile. Rosita will be on the left.

Brown's Spring was cattle country in 1870. Daniel Baker was a cowboy who had picked up some galena samples and put them on his windowsills to glitter in the sunlight. Dick Irwin, for whom the town of Irwin was named, had also been in the area with his friend Jasper Brown in June of 1870. They had come prospecting and collecting samples. When they had their samples assayed, the assayer claimed the ore was not worth processing. Dick Irwin kept looking until finding his strike. The town of Irwin, Colorado, was named after him. One story is that Dick Irwin named the town Rosita for all the wild roses growing in the area. In another story, a heartbroken Frenchman had come out to the area after losing his beloved and named the place Rosita because it was so beautiful.

In 1872 prospectors rambled through the area, dug holes, and tried some placer mining. The Hardscrable Mining District was formed. In April 1873, a small camp began to form around a cluster of tents, a few cabins, and a boardinghouse.

In 1874, the Pocahontas Mine was found, followed by the Humbolt Mine. These were Rosita's most productive mines, and the town soon grew into one of Colorado's biggest cities. The main street was called Tyndal Street. There were saloons, The Snowy Range Hotel, The Windsor House Hotel, some less well-known hotels, restaurants, and many other businesses. In 1874 Rosita had the biggest brewery in Colorado. Rosita also had a dairy producing cheese, 250 pounds daily, until the cows ate a patch of wild garlic, which contaminated their milk. Rosita was one of the first towns in Colorado to get telephone service.

By the mid 1870's, the population reached 2,000. About this time, there was a serious problem with overlapping claims and claim jumpers. A man named Colonel Boyd moved to Rosita from Denver. He was not a real colonel, but he had a lot of money and simply appropriated that title. Within a few weeks, a Denver banker named Stewart moved to Rosita.

Soon, the phony Colonel and the erstwhile banker collaborated in an effort to get the Pocahontas Mine, which was owned by Mr. Herr. Boyd and Stewart paid all of the mine employees to leave town, and put their own men in place with guns. Their foreman was an escapee named Major Graham, who was not a real major. When Mr. Herr approached his mine, he was threatened with a gun and treated as a trespasser. Mr. Herr and all the residents of Rosita were shocked. The next day Rosita's town constable got arrest warrants, assembled a posse, and went to the mine. They arrested the claim jumpers and took them to court. Stewart soon paid the fines, and the men were freed. The men in turn, marched right back up the hill, and took over the Pocahontas again. For several days the interlopers terrorized the town, threatening everyone. No one went to work, and all of the businesses stayed closed. On the second day, a vigilante band and armed citizens congregated in the middle of town. Graham and his men had been watching them and around 1:00 a.m. Graham and his men attacked first. It was a war, a gunfight that went on until dawn, when Graham's men returned to the mine. The vigilantes met them, and Graham was shot to death. The rest of the men were captured. Mr. Herr reclaimed his mine. That was the last time a claim was jumped in Rosita. The vigilantes threw Graham's body into an abandoned mine hole.

The Pocahontas, the Senator, and the Humbolt, the top mines in Rosita, produced between one and two million dollars in silver.

At the center of community life were the lodges: the Masonic Lodge established in 1879, Knights of Pythias, and the Odd Fellows. These lodges held special events for their members, and established saloons and dancehalls.

Commodore Stephen Decatur, for whom the town of Decatur was named, spent his last years in Rosita, and is buried in the cemetery there. He had been a Commissioner for Colorado in Philadelphia's 1876 Centennial Exposition. When Colorado's appropriated funds ran out a month and a half before the fair ended, the Commodore spent everything he had to maintain Colorado's Mining Exhibit. He was able to keep the exhibit open until the end of the fair, but this act of generosity essentially bankrupted him. He was never reimbursed. The Commodore was destitute. He then lived in Silver Cliff for a while until he was disabled in an accident. He then drifted over to Rosita, where someone gave him a place to live, and the townspeople cared for him. They made sure he always had a warm meal and clean clothes. He would tell the children wonderful stories about history, mining, cowboys, and Indians. He died in the summer of 1887.

The Colorado State Directory for 1878 listed the following on Rosita:

ROSITA
County seat of Custer County, center of the Hardscrable Mining District. Daily mail and express. Daily stages to Canon City, triweekly to Pueblo, semi-weekly to Colfax and Ula. Population 1,200. Distances: To Canon City, 30 miles; To Pueblo, 50.

ROSITA DIRECTORY
Town Trustees-. E. Austin, Charles F. Blossom, Moses Bloucett, James A. Melvin, Alexander Thornton.
Town Clerk- J. W. Brewster.
Postmaster- Thomasson.
School Board- J. W. Warner, H. Tucker, Wm. McLaughlin.
Public School- H. C. Langley, principal.
Justice of the Peace- H. Tucker, J. C. Duncan.
Churches.
    Catholic- Fr. Pinto, priest.
    Episcopal-
    Methodist- Rev. H. C. Langley, pastor.
    Presbyterian- Rev. Teitsworth, pastor.
Society. Rosita Lodge, No. 21, I. O. O. F.

BUSINESS DIRECTORY
Agricultural Implements. Webb & Tompkins.
Assayers. Braun, Theo. F. Beaumont & Co.
Attorneys. Adams, Geo. S. Buell Jas. Offenbacher, W. A. Warner, J. W.
Bank. Bank of Rosita.
Bakery. Wiley, Robt.
Billiards. Austin, Eddie. Kohn, James. Slavick, Louis. Smith, E. P.
Blacksmiths. Backman, S. J. Neely & Sanders. Sorter & Co.
Barber. Work, Henry T.
Boots and Shoes. Diehle & Co. Hopkins & Co.
Bowling Alley. Gilliam, Charles.
Brewery. Townsend, O. P.
Builders. Barret, W. E. Duncan & Franz. Nelson, Charles. Sargent, W.
Cigars and Tobacco. Payne, B. D.

Clothing. Slavick, Louis. Smith, B. F.
Conveyancers. Brewster, J. W. Thomasson, Tower. Tucker, H.
Crockery. Webb & Tomkins.
Dry Goods and Groceries. Austin, H. E. Miller, F. L. Wilson, J. A.
Dressmaking. Blush & Baker, Mrs. Tolle, Miss
Drugs. Kent & Callahan.
Express. Miller, F. L.
Fruit. Miller, F. L. Payne, B. D.
General Merchandise. Austin, H. E. Miller, F. L. Miller, D. D. Rice, James. Furniture. Bartlett, A. Keller, Joseph.
Grain and Feed. Blush, Daniel.
Groceries. Miller, F. L. McElhinney, A. M.
Hardware. Webb & Tomkins.
Harness and Saddlery. Daniels and Robert.
Hotels. Grand View, R. N. Daniels. Humbolt, V. B. Hoyt. Melvin House, J. W. Brewster. Snowy Range, G. Nodine.
Jewelry. Pagand, W. H.
Laundry. Sing Lee.
Liquor. Austin, H. E.
Livery. Gorman, Hudson & Sieber.
Lumber. Creager, J. C. Edwards Bros.
Millinery. Tolle, Miss.
Mining Agent. Brewster, J. W. Thomasson, Tower.
Mining Tools. Webb & Tomkins.
Meat Markets. Hudson & Sieber. Knight, W. L.
Musicians. Cropsey, Mrs. Allen. Rich, John.
Newsdealer. Rice, James.
Newspaper. *Rosita Index*, weekly, Charles Baker, Prop.
Physicians. Blatchley, F. C. Camp, W. S. Blake, O. F. Hoge, J. M.
Powder. Webb, N. H.
Reduction Works. Rosita Reduct`n Works. Pennsylvania Reduction Works. Prof. Baugh`s Reduction Works.
Restaurant. Youman & Co.
Saloons. Austin, H. E. Kohn, James. Smith, E. P. Slavick, Louis.
Saw Mills. Apperson, Wm. Dinning, James. Hager & Co.
Sign Painting. McNeff,-
Stoves and Tinware. Webb & Tomkins.
Surveyors. Holmes, W. H. Wulsten, Carl.
Wagon Maker. Finn,-

Less than ten years later, the town of Rosita had seen its population fall by about 75% to 300 and had fewer stores and conveniences. *The Colorado State Directory* for the year 1896 lists the following for Rosita:

ROSITA

Town in Custer county. Daily mails to Silver Cliff, 7 miles and Westcliff, 10 miles. Population 300. Altitude 8,565 feet.

Ester, John, mining
Fabian, J. F groceries
Fisher, Thos, county assessor
Fisher, Mrs. Thos, postmistress
Husted, H. transfer
Ingraham, Fred, Livery
Kennedy, Mrs. Wm, hotel
McKinnon, Jas, stock dealer
Myers, John, stock dealer
Payne, B. D., groceries
Pringle, Jas, mining
Schmier, Mis R, dressmaker
Schoolfield, W. A., drugs
Schriver, Henry, mining
Schwabe, J. L., stock dealer
Simmons, Reuben, physician.

On 10 March 1881 a fire destroyed four blocks of the business district. The town was never completely rebuilt. The mines in the area started drying up.

One night in late January 1884, there was a lynching in Rosita. Orion Kurtz, the bouncer at the Odd Fellows Saloon, cut off a couple of drunks, Frank Williams and John Gray. Williams and Gray were enraged. They shot and killed the bouncer and were thrown in jail. The townspeople were stunned. The next day, unsigned invitations to a "Necktie Party at midnight" were distributed in Rosita and neighboring Silver Cliff. Vigilantes stormed the Rosita jail, took the two men, and hanged them outside the jail. The lynching was reported without comment in *The White Pine Cone*, the local newspaper, on 1 February 1884.

In 1884 the Bassick Mine in neighboring Querida shut down, forecasting the slow death of Rosita. Within six years, the town was empty.

In the mid 1950's Metro Goldwyn Mayer filmed a movie in Rosita. The film studio changed Rosita's remaining buildings and moved them around to create a "realistic" film set.

Today, little is left of what was once one of the largest cities in Colorado.

FULFORD   (Nolan's Creek Camp, Camp Fulford)
Eagle County

Directions: From the town of Eagle, go south on Brush Creek Road, which is also Forest Route 400, for about ten miles until you reach Forest Route 415. Take this road east for about six miles. You will pass a campground, and the road turns into Forest Route 418. After about two miles and six meandering switchbacks, you will cross Nolan's Creek. Fulford is immediately after the creek.

The area is known for the legend of a lost cache of gold, buried by an avalanche somewhere down the craggy slopes of Slate Mountain, and Buck Rogers, the man who spent the rest of his life searching for his lost companions and the treasure they had wrested from the mountain. His tale of lost gold and tragedy was recounted in *The Denver Republican* on 8 May 1892.

It was in 1849, three years after the ill-fated Donner party left Springfield, Illinois, that Buck Rogers recruited a party of about 30 prospectors from Bloomington, Illinois, and neighboring towns. The "49ers" headed West, to seek their fortunes in the newly-discovered gold fields of California. By October, the weary travelers reached the placer fields of Pike's Peak. Ignoring the threat of early mountain snows, the men pressed on, lured by dreams of a gold strike in California. They had traveled northwest for eight days, about 170 miles from Pike's Peak, when some of the men saw the unmistakable glint of gold in a small stream. Buck Rogers and five of his companions thought they had discovered the gold they were seeking and set up camp along the stream.

The rest of the band continued on to California. Winter would soon be upon them, and they had a long way to go.

Buck Rogers and his party stayed at the campsite to pan out some of the gold. In addition to the gold dust that sparkled in their pans, they found large nuggets just below the surface. Encouraged, they started digging, and after only six weeks they struck a lode of rich ore. They stashed their gold in a drift. In a short time, they had accumulated between $60,000 and $100,000 in gold. Winter snows began falling, and the men were nearly out of supplies. They drew lots to decide who would make the 150-mile trek to the nearest camp to get supplies. Rogers was chosen to go. He started out in a snowstorm on the morning of November 26, carrying a buckskin bag with $500 in gold. The snowstorm whipped and swirled into a blizzard. It took him seven days to make it to the nearest camp. Rogers waited out the winter storms in the safety of the local saloon where he soon spent all the gold intended for supplies. Time passed.

Rogers was stranded in camp until January 12, when he started back to his campsite, worried about his companions. It took him only five days to reach the campsite, but everything was different. The 49ers' cabin and diggings were gone. The mountainside had been fractured and sheared off, exposing the barren grey slate. An avalanche had carried off all traces of his five friends, their campsite, and their treasure. For days he frantically dug and searched for his entombed companions. The five prospectors were carried away and buried in the debris of the avalanche somewhere along the base of Slate Mountain. They were never found.

Rogers was a broken man, burdened by the guilt that his companions waited in vain for his return and were killed while he drank and socialized in a warm saloon. He wandered from camp to camp, from saloon to saloon, telling his tale of guilt and lost treasure to anyone who would listen.

For decades, prospectors and treasure hunters searched for the remains of Buck Rogers' companions and the gold buried at the base of Slate Mountain.

In 1887, some rich gold strikes were made along Brush Creek in Eagle County. Almost four decades after Buck Rogers, a prospector named Nolan was the first to rediscover gold in the area. His discovery triggered a gold rush, and the little camp that formed was called Nolan's Creek Camp. Soon, more gold strikes were made. Nolan himself did not live to see the prosperity he had started. One day Nolan was walking across Brush Creek carelessly carrying his shotgun, when he tripped, the gun discharged, and he was fatally shot in the head. The stretch of Brush Creek where he died was renamed Nolan Creek.

Nolan's Creek Camp continued to grow and soon boasted 300 residents. The town was divided into an upper and a lower part of town. Up Brush Creek was the Fulford Ranch, home of an early ranching family in Eagle County. Their son, Arthur H. Fulford, operated a stagecoach station on the Fulford Ranch at the mouth of Brush Creek, not far from Nolan's Creek Camp.

In 1891, James Fulford met an old prospector with a strange tale. When old Buck Rogers died in 1881, he had left the prospector his journal. The prospector spent the next ten years searching for the location of Slate Mountain and its lost cache of gold. During one of his many treks into the mountains, the old prospector was stunned to see that the mountainside he was traversing was flat and avalanche scarred. A small mound swelled up at the base of the mountain. This, he realized, had to be the legendary Slate Mountain. For months he dug test holes and searched for clues. He finally found an abandoned tunnel and so much gold that he decided he needed a partner. He returned to town and asked James Fulford to be his partner. Fulford and the prospector prepared to return to Slate Mountain. Two weeks later, the prospector was killed in a drunken brawl in a Red Cliff gambling house.

With his partner dead, Fulford set out to find the prospector's cabin, hoping that he would find clues to Slate Mountain and the lost treasure. If he was very lucky, he might find old Rogers' journal. He was more than lucky. He found the cabin, the journal, and knew that he would eventually find the lost treasure of Slate Mountain. In midwinter, Fulford was on his way to Aspen when he was caught in a snow slide on New York Mountain. His body was never recovered.

The treasure of Slate Mountain was again lost.

On 4 January 1892, *The Denver Republican* reported that Arthur H. Fulford was killed in a snowslide. This is likely some confusion, and it was probably James Fulford who was killed. On December 31, 1891, New Year's Eve, Arthur H. [or, James] Fulford went to visit some property about 15 miles from his home in Eagle. When he did not return, his family organized a search. Fulford's family traced his footprints in the snow until the footsteps were obliterated in a massive snowslide estimated to be a mile wide, two and a half miles long, and 75 to 100 feet deep. He left a widow and three children. Later search parties were never able to recover his body.

The men living at Nolan's Creek Camp began calling the camp, Camp Fulford, in memory of their lost friend. As the camp grew, the townspeople shortened the name to simply Fulford. The upper section of Fulford had saloons, boarding houses, hotel, and stables; the school and businesses were in the lower section. Cabins were scattered all over. Fulford peaked in the 1890s. The districts mines had attracted the attention of the Denver newspapers, bringing more prospectors to the area, who staked over 500 claims. Not many were good enough to be developed like the Cave, and the Polar Star, which were the best ones, and then there were the New York, Kittie B., Layton, Adelaide, and the Mendota. Throughout the 1890's, gold, copper, and lead were mined and transported by wagon to the nearest railroad station in Eagle. A few mines like the Cave could be accessed only by climbing up a cable to a cave opening on a high cliff. The famous Fulford cave is only a mile south of Fulford, with a campsite right next to it, and is shown on many maps.

All of the mines in the area played out around 1910. The town was dying, too, until silver was discovered. Fulford enjoyed another boom until 1918, when the silver ran out. After 1918, the town was deserted for good. Just east of lower Fulford about a mile was the camp called New York Cabins. Just east of upper Fulford about a mile was the camp of Adelaide Park, and as Fulford grew, it absorbed both of these camps. When Fulford was finally abandoned, so were New York Cabins and Adelaide Park.

Today, scattered ruins mark the area.

HOLY CROSS CITY
Eagle County

Directions: Holy Cross City is four miles northwest of Gold Park, and about 12 miles southwest of Redcliff at an elevation of 11,335 feet. Driving south on U.S. Highway 24, after passing the exit for Redcliff, about three miles farther south on U.S. Highway 24 is Forest Route 703 on the right. There will be a sign pointing down this road, to Gold Park. Follow this road for 5 ½ miles. At this point turn right onto Forest Route 759. This is a very rough jeep trail. Follow this northwest for about two miles, then southeast for about a half mile, then southwest for about a half mile, and you are there.

In a beautiful meadow almost at the timberline, and adjacent to Mount Whitney, the town of Holy Cross City was settled in 1880. Holy Cross City was named for the spectacular natural cross image that is imbedded in the mountain a little more than three miles north of town. The actual cross image is a snow-filled crevice, making the cross a snow white against the dark mountain. Holy Cross came to life about the same time that Leadville was booming. Leadville was located 37 miles away. The whole area was filled with prospectors, miners, and fortune hunters, and soon rich discoveries were reported in the area. Homestake Creek had been found to be full of placer gold. French Peak and Homestake Mountain had been found to be rich in gold also. These areas had been completely staked off in mining claims. About a mile directly above Holy Cross City was the Pelican Mine with its two tunnels. The Comstock Mine was also a good producer, and the Molly Mines were good too, with its four tunnels. The deepest of these was almost 300 feet. Some of the other, less productive, mines were the Belle of the West, the Bachus, the Hunkidori, and the Tip Top. These mines played out by early 1883.

By 1882, *The Colorado Business Directory* reported the following:

HOLY CROSS CITY
Mining camp in Summit County, 4 miles from Gold Park.

Durbin Bros., general mdse.
Farquher, John, justice of peace.
Frey, R., general merchandise.
Morrow, R.W., saloon.
Park, Mrs. M. E., boarding house.

By 1883, according to *The Colorado Business Directory,* the number of businesses had tripled. The population grew from 200 to 300. One residential main street was lined with homes. To the south was a street with a bridge that turned into a walking path leading uphill. On the east side of town was the business district with the Timberline hotel and other businesses. One boardinghouse had more than 20 rooms. A network of roads led from the town to the many mining properties. There were a few mining companies, but the Gold Park Mining Company controlled most of the mining properties.

When the Treasure Vault Mill was built, the construction workers did not use the main road into town, a road that was owned by the Gold Park Mining Company. Everything had to be hauled by ropes and pulleys up a dangerous cliff and over Fancy Pass, using manpower and a lot of donkeys. When everything was finally in place and built, the mill manager mysteriously disappeared. The mill faced a series of problems and closed down in less than a month.

The Holy Cross School was established in 1883, the same year that daily mail and stagecoach service was inaugurated. By the end of 1883, the mines in the area had all played out. They all closed down, one after the other.

The area is beautiful, but the only reason for the town was the mines. The weather was always cold, the roads were very bad, and the town was remote. After the mines played out the residents left.

In the late 1890`s, the Holy Cross Gold Mining and Milling Company tried to revive the mining and the town, bringing in workers, electricity, and even a phone line. Unfortunately, they were unsuccessful, and the town was empty again by 1899.

The army built a training base at nearby Camp Hale, and during World War II, Holy Cross City was located within the U.S. Army bombing range and training facility. Camp Hale now consists of a field off U.S. Highway 24 with scattered cement foundations, and marked by historical interest markers. As recently as the year 2000, traces of live ammunition and an unexploded rifle grenade were found in the area, and hikers are warned to be cautious of any suspicious devices.

Today, Holy Cross City consists of a few cabin ruins and about 20 foundations.

The Legend of Holy Cross

In 1542, a party of Spanish Explorers crossed through the Colorado Rockies, led by their leader DeSoto. According to legend, a Franciscan friar who accompanied the explorers somehow angered one of DeSoto's soldiers, who then stripped the friar of his crucifix, habit, and clothing. The soldier sent the friar alone into the wilderness. This was a death sentence for the unarmed friar who would be unable to protect himself against the mountain lions, bears, and Ute Indians. A Chickasaw brave had been quietly following the party of explorers and witnessed everything. After the friar was left alone, the Indian approached the dejected friar. Seeing the Indian, the friar kneeled down and, expecting his imminent death, drew a cross in the dirt. The Indian took the friar back to his tribe, and took him in. The friar was then living, hunting, and exploring with the Indians, until one day by chance the friar came upon part of Desoto's Band of explorers. He recognized the soldier who had taken his clothes and his crucifix, and cast him out to die. The soldier was now wearing the friar's habit, and carrying his crucifix. Enraged, the friar drew his hunting knife and killed the disguised soldier. Realizing what he had done, the friar was afraid and too ashamed to touch his crucifix. He left the Indian tribe that had taken him in. He needed to be alone to seek forgiveness. He climbed the mountain as high as he could go above the clouds. Finally, he kneeled down and prayed to God for forgiveness. As he prayed, the clouds cloaking the mountain peak slowly dissipated, revealing a white cross on the dark face of the mountain. . . a symbol of the mercy of God.

AMERICAN CITY
Gilpin County

Directions: Coming from Back Hawk on State Highway 119, go north for about a mile and a half. Then on your left will be County Road 4, also called Apex Road. Take this for about five miles. You will then be at the town site of Apex, on the main street, in the middle of town. County Road 4 takes a left turn here, crossing a very small seasonal stream. Follow this road for about a mile until you are on the mesa of Idaho Hill. You will see the soaring peaks of the Continental Divide here, with James Peak directly ahead. At this point you will need to take an unmarked four-wheel drive trail on your right. Follow this for a mile to the site of American City. There is a very small sign noting that this is American City; it is private property so you will need to get permission. The trail was in bad shape when the authors saw it in the summer of 2001, but you should be able to make it with a high clearance or four-wheel drive vehicle.

In the late 1890`s, the community of American City sprang up on the side of Colorado Mountain. It was about a mile and a half north of the smaller town of Nugget and about a mile and a half north west of the larger town of Apex. These towns were friendly neighbors, sharing their mines, mills, and businesses. The main mining company in town was the Boston Occidental Mining and Milling Company. The largest structure in town was the three-story Hotel Del Monte, an upscale, popular place to stay, which offered spectacular views of the valley below. Today, as you enter American City, your first sight of the town will be the intact ruins of Hotel Del Monte on the left. The American City School was graciously shared with the children of Nugget. The road through Apex ends at the mine, which has now been cemented over and resembles a square-bottomed swimming pool. There was no post office in town, so the residents used the Apex Post Office.

The life of American City was brief. After the turn of the century, the gold in the area mines started to play out. By 1905, the town was abandoned.

In 1911 an eastern movie studio filmed on location in American City. Reportedly, they did some work here, changing and moving around some of the original buildings. A few summer cabins were built. Today, it is hard to distinguish American City ruins from the movie set and later structures. There are a few remains left.

APEX
Gilpin County

Directions: Coming from Black Hawk, take State Highway north for about a mile and a half, on your left will be County Road 4, which is also Apex Road. Take this for about five miles, and you will be at Apex.

In the middle of the town site, directly across from what was the town hall, is a road to the left. About 50 feet up this road, on the right side, are a few old cabins dating to the 1890's. Between these old cabins is a very old, overgrown, narrow road hidden in the heavy forest. The road is so overgrown it appears to be a driveway, but it is actually a road leading to a few dozen old homes and other structures, with an old church at the very end.

The town of Apex sprang up in the 1890's in the middle of the Pine Creek Mining District with the discovery of rich gold and silver deposits. This is not to be confused with the other town of Apex that once existed between Golden and Morrison. The Gilpin County town of Apex is located in a beautiful mountain forest, and the grassy meadows are filled with wild raspberries, strawberries, and sweet peas. The little town of Nugget was a mile and a half to the west, American City was a mile and a half to the northwest, and Central City was about seven miles southeast of Apex.

Apex had numerous mines -- the Annie Mascot, Evergreen, Jersey City, Rooster, Schultz Wonder, Tip Top, Wetstein, and Yellow Medicine-- but the most valued was the Mackey Mine. Dick Mackey discovered it in the 1870`s, and it went through many owners. A man named Mountz was a member of the group who finally owned the mine. The men had begun to work the mine, and found only mediocre, very low-grade ore. Mountz's business partners disappeared in the middle of the night with about $30,000 – nearly all their money. They left Mountz with $400, possibly an oversight or perhaps so that he would not be totally destitute. Mountz was distraught, angry, and betrayed. He bought all of the dynamite that he could afford and blew up the mine. The next day when the dirt and dust cleared, Mountz went out to take a look at his shattered dreams and discovered the biggest, richest vein of gold ore that he had ever seen. It assayed at $1,800 per ton. He immediately ordered a couple of horse teams and wagons from Central City, and from Denver he ordered a stack of ore bags, all on credit. Mountz was left a very rich and happy man, and the sole owner of a very productive mine.

During its brief life, Apex was a busy, prosperous town. The population peaked at around a thousand, although there was likely a large transient population of miners and prospectors drifting through the area. It boasted its own newspaper, *The Apex Pine Cone*, and over 100 different businesses. It had two hotels, the 22-room Apex Hotel and the Pine Creek Hotel, and a busy social center, the multi-faceted Pioneer Hall, which also served as a dance hall and church. Hundreds of cabins were scattered in the area. After the turn of the century the mines stopped producing and could no longer support a town.

Winter snow was so deep in Apex that some of the creative residents built two-story outhouses to accommodate the drifts.

The Colorado State Directory for 1896 listed the following on Apex:

APEX
Growing mining town in Gilpin County, 8 miles northeast from Central City. Two daily stages.

Apex Hotel, P A Richards, prop.
Brown, F. W, barber.
Converse, Geo. A, gen. Mdse and postmaster.
Fish, Chas. A. mail carrier.
Gableman & Co., H, groc`rs.
Kemp, S W, min`g engineer.
Parkinson, J B. saloon.
Pine Creek Hotel, Jos. Moody, prop.
Schneider, G W, min`g en`gr.
Scott & Clow, saloon.
Wilson, W D, shoemaker.

By 1900, there were only 237 people in Apex, and ten years later, the population had fallen to 63. By 1918, all the businesses had closed except the Apex Mercantile and the post office. By 1920, there were only 49 people in the struggling community.

In 1939, a raging fire ravaged buildings on the west side of Gilpin Avenue, the main street.

In December 1947, four Denver entrepreneurs, Ralph Balant, Bill Goe, and Donald and Horace Hix, built a ski area on the slopes of Idaho Mountain edging Apex. They bought eight old Apex mining claims that stretched across the Arapaho National Forest. The ski area had two runs and a 1,000-foot rope tow. The enterprise was short-lived and was unable to breathe new life into Apex.

The residents moved away, although a few people still come and spend time here in the summer. This is a very beautiful area. The old store still stands, and a few of the old cabins are used as summer weekend getaways.

An old storefront in downtown Apex.

BALTIMORE
Gilpin County

Directions: Coming from Rollinsville on State Highway 119, turn west on County Road 16, which follows Boulder Creek and the train tracks. Take this road for 4 miles and then on the left will be the trail that goes south down Jenny Lind Gulch. Take this trail for about a third of a mile and you will be at the site of Baltimore.

The little town of Baltimore was founded four miles east of the Moffat Tunnel and four miles west of Rollinsville at the mouth of Jenny Lind Gulch. The Gulch was named after the famous Jenny Lind, a beautiful theatrical entertainer of the 1880's. Men throughout the West were in love with Jenny Lind.

Baltimore formed in the early 1880's after gold was discovered. The population eventually peaked at 300. All of the usual businesses graced Baltimore's main street: a two-story hotel, a saloon, and a general store. Baltimore was well known for its upscale little social club. There was an opera house with beautiful scenery and heavy curtains.

Only a few cabins are left of Baltimore, and you have to look hard to find any trace of a town. A couple of cabins in the area are used in the summer. Today, the site of Baltimore is on private property.

EAST PORTAL
Gilpin County

Directions: Coming from Rollinsville on State Highway 119, you will see County Road 16 heading west, following Boulder Creek and the railroad tracks. Take this for eight miles until the road dead ends. This is East Portal and the Moffat Tunnel.

East Portal was a little town formed by the men building the Rollins Pass (also called Corona Pass) Road and railroad grade in 1901, which was completed in 1909. It was named for David H. Moffat, who designed and built it. Unfortunately the pass was very dangerous because of the heavy snows and bad weather at such a high altitude of 11,671 feet. The railroad that used the pass was the Denver & Salt Lake Railroad. The railroad company spent about 41% of its budget to shovel the snow off the tracks. Huge snow sheds were eventually destroyed by the snow. David H. Moffat dreamed of building a tunnel through the Continental Divide to save time and the burdensome expense of keeping the snow under control, but the Rollins Pass Road had cost him and his partners most of their money, and they could no longer afford to build another road. He then pushed for state funding to build the tunnel. The state agreed, and the Moffat Tunnel project started in 1923. At a cost of $18 million, the tunnel was finished in 1927.

It was during this time that the town of East Portal thrived. The townspeople left after the tunnel was finished.

Directly left of the tunnel is a hiking path, which is well marked and maintained by the state. By following this path you will pass the last remains of East Portal, consisting of about a dozen cabins.

The Moffat Tunnel, named for David H. Moffat, was started in 1923 and finished in 1927 at a cost of $18 million. It cut through the Continental Divide to open up transportation. The little town of East Portal existed during the brief period of construction.

NEVADAVILLE (Bald Mountain, Nevada City)
Gilpin County

Directions: From Central City, go southwest about a mile on the Virginia Canyon Road, to Nevadaville.

One of the most easily accessible remnants of the Colorado gold rush, Nevadaville began as a mining camp in 1859 with the discovery of the famous Burrows lode. It crowned a trail of gold mining towns - Black Hawk, Mountain City, Central City, and finally Nevadaville - that stretched along a gulch and reached up the side of Bald Mountain. The four separate towns were nearly indistinguishable, bumping up against each other and jostling each other, competing for resources.

The little boomtown was first called Nevada City, but the post office soon changed the name to Bald Mountain, reportedly to prevent confusion with another town named Nevada City. This is how the town was listed in the 1870 Federal Census. The original Nevada City, if it ever really existed, has not been found. The miners rejected the new name for their community, and found a compromise in the name Nevadaville. The Post Office eventually gave in, and in the 1880 Federal Census the town was called Nevadaville.

Nevadaville was formally incorporated 7 July 1870.

Numerous mines were sited within half a mile of the center of town. The most important were: the American Flag, the Burrows, Camp Grove, the California, the Dyke, the Flack, the Gardner, the Hubert, the Indiana, the Jones, the Kansas, the Mercer County, the Missouri, the Prize, the Rising Sun, and the Sunderberg. At one time, Nevadaville had 175 stamp mills in operation, with more idle because of a lack of water.

By 1888, Nevadaville had a population of 1,060, a number of thriving businesses, a doctor, a postmaster, a town clerk, and five saloons. A public school was built at a cost of $3,000.

The 14th annual *Colorado Business Directory* published in 1888 by James R. Ives included a half-page business directory for Bald Mountain:

Ashbaugh, A., physician
Brown, Geo., barber
Brum, August, blacksmith

Comer, D. B. druggist
Combs, Robert, carpenter
Cook, T. P., paster, M. E. Church
Ellis, Pasco, saloon
Grenfell, J. H., confectionery
Hicks, J. R., shoemaker
Hocking, Joseph, groceries
Koby Bros. Groceries
Lawry, Thos. H., saloon
McGonigal, Daniel, postmaster
Nankervis & Rhem, meat market
Nankervis, Wm., saloon
Nichols, Wm., mayor
Noble Bros., meat market
Parsons, I. M., livery
Rachofsky, H., dry goods & clothing
Rachofsky & Lonn, blacksmiths
Ratliff, J. W., justice of the peace
Richards, J. C., town clerk
Rowe, Geo., tobacco and cigars
Sparks, John, carpenter & millright
Stedman, M., saloon
Trembath, Jas., saloon
Trenoweth, C. clothing

Remnants of Nevadaville still standing on the main street include the firehouse, a few saloons, a store, and a few homes. Some of these buildings have been renovated as tourist stops. Many foundations remain between Nevadaville and the outskirts of Central City. Tailings piles litter the area.

NUGGET
Gilpin County

Directions: Coming from Black Hawk take State Highway 119 north for about a mile and a half. On your left will be County Road 4, take this for about five miles. At this point you will be on the main street in the middle of what was once Apex. County Road 4 turns left here and goes uphill for a short distance. The road is a little rocky with a few potholes here and there, and you should have a have a four-wheel drive vehicle. Keep following the road for a little over a mile until you are on top of a hill looking down on a beautiful valley. Start to drive down the hill; about two-thirds of the way down, there will be a mine on your left and a road. Take this road, and it will fork almost immediately. Take the right fork; the left fork leads to the mine and that road is chained off. The right fork is rough, but take it a few hundred feet until you can find a spot to turn around and park. This is the site of Nugget.

In the mid 1890's, about a mile west of the town of Apex, the little gold mining town of Nugget came to life. The town was on the west side of Idaho Hill. Nugget had one mine that supported the town. The oldest opening of the mine is towards the top of the hill. Another opening to the mine is about 100 feet lower on the hill, boarded up with wooden doors. Beautiful pyrite specimens are scattered all over. A lot of old mining machinery has been left behind here. This was part of the Pine Creek Mining District. The Nugget post office was established in 1895. Gustav Meyer was the postmaster and ran the general store. The school children in Nugget had to go to school in American City, a few miles away. Between American City, Apex, and Nugget, cabins were scattered everywhere.

By 1901, the Nugget Mine that supported the town slowed down, the town's residents moved away, and the post office closed. Nugget was and is a very isolated location, although it is beautiful. After the townspeople abandoned Nugget, the postmaster and store clerk, Gustav Meyer, reportedly lost his sense of reality and died in the Pueblo Mental Institution.

Almost nothing is left of Nugget now, and one old cabin marks the road that you turn down. The mine will be above, on the hillside to the left. To the right the hillside slopes gently down to the bottom of the valley, which is filled with bushes and a little stream. In the winter a thick blanket of snow completely covers the valley, hiding it for most of the year. In the summer, a few all-terrain vehicles, and four-wheelers come through the area. You can see the old Mackay 444 mine and cabin if you walk farther down the road. The mine is barely recognizable as a large old building that has fallen down on itself, next to a few tailings piles on the right.

TOLLAND (Mammoth)
Gilpin County

Directions: Coming from Rollinsville on State Highway 119 turn west on County Road 16 and continue on this road for 4 ½ miles and the site of Tolland will be on your left.

Tolland was founded as a stagecoach station in the late 1800`s for the coaches that traveled between Rollinsville and East Portal. When David H. Moffat built his railroad over Rollins Pass, Tolland became an important railroad station. Soon a little town grew up here. Tourists stopped to enjoy the beautiful scenery and the fresh air. After the Moffat Tunnel was built, even more tourists came through the area. The town was originally called Mammoth, but Mrs. Charles B. Toll renamed it Tolland. She owned and operated a hotel here in the early 1900`s. Reportedly, a few thousand tourists came through town every day.

It was discontinued as a railroad stop, and then bad weather and a fire claimed the small town. Less than a dozen cabins remain.

WIDE AWAKE
Gilpin County

Directions: Coming from Black Hawk, take State Highway 119 north for about 2 ½ miles. At that point you will see a sign marking Pickle Gulch, which is on County Road 15, on your left. Take this road for about two more miles, and the site of Wide Awake will be on your left side of the road.

In the 1860's, at the end of Missouri Creek, not far from Missouri Lakes, the little town of Wide Awake was founded. It prospered and grew from a tent city into a bustling mining camp with 500 residents. Little information is available on this little town. The five area mines played out, and the little town was abandoned in the 1920's when the mill shut down.

Wide Awake's last resident, Wallace Stevens, died in 1964 at the age of 97. According to legend, Stevens had buried $20,000 in gold in a gunnysack in a hole somewhere around Wide Awake. As he was dying, he told a friend to "look in the hole we dug." Wallace dug many prospecting holes, and the one with the buried sack of gold has not yet been found. Unless this was simply the delirious rambling of a dying man, there is still a bag of gold waiting to be found.

The abandoned town site is quiet now except for the whispers, shrieks, and noisy pranks of its Tommyknockers, those mischievous Cornish elves that came to the new world with Cornish miners.

In the spring of 2001 it appeared that there was some active mining.

The abandoned town site of Wide Awake, quiet now except for the whispers, shrieks, and noisy pranks of its Tommyknockers, those mischievous Cornish elves that came to the new world with Cornish miners.

ALPINE STATION
Gunnison County

Directions: Alpine Station was at the west entrance of the Alpine Tunnel, eight miles northeast from Pitkin.

Alpine Station was founded in 1880 by the railroad workers who lived there for the five years it took to build the Alpine Tunnel. The men lived in a stone bunkhouse, and there were other stone buildings here. Brutal winters brought blizzards and avalanches, which often derailed and stranded trains happened often. In March 1884, an avalanche, reportedly triggered by the noise from a train, carried the little camp of Woodstock to the bottom of the mountain. Woodstock was located below Alpine Station. Only three people of 17 survived.

A small fire burned part of Alpine Station in 1906. The town was rebuilt, but when the Alpine Tunnel collapsed in 1910 killing several people, everything was boarded up and closed.

As of August 2000, the old railroad grade from the site of Hancock to Alpine Station, was only a trail. The tracks had been taken out long ago. At the beginning of the trailhead, people were camping, four-wheeling, and driving all-terrain vehicles. Motorized vehicles are not allowed on this trail. This is a wonderful hiking trail and camping spot.

BALDWIN
Gunnison County

Directions: Baldwin is 17 miles northwest of Gunnison. Leaving Gunnison on State Highway 135, take Ohio Creek Road to Baldwin.

Baldwin's Post Office was established in 1883, in a little community south of Baldwin. This was cattle country until 1896 when someone struck gold, igniting a small boom. When the gold ran out, the men started to mine coal from the vast deposits around Baldwin. Baldwin became one of the top coal camps in Colorado.

When the coal miners found a more productive site on the other side of the hill, the railroad moved its tracks to accommodate them. With the mine and the railroad on the other side of the hill, the miners soon moved the entire town.

Mining continued in the area until 1950. Everyone left except for one man, Joseph "Peanuts" Berta, who remained until his death in 1967. He was buried across from town.

BOWERMAN
Gunnison County

Directions: Bowerman is at the base of Wuanita Pass, just three miles north of the Waunita Hot Springs site and three miles south of Pitkin.

Bowerman was built on rumors of a gold strike that never happened. J.C. Bowerman was a prospector who lived with his wife in a cabin at the base of Waunita Pass. The trouble all started when he gave his wife some gold nuggets. Bowerman's wife enjoyed the sparkling "pretties" and frequently carried them with her as a conversation piece. Mrs. Bowerman was a sociable woman, who must have found it difficult to live in a cabin in such an isolated area. One can imagine how Mrs. Bowerman enjoyed going to town to get supplies and chat with storekeepers and other shoppers. During one such trip to town, Mrs. Bowerman showed her glittering gold nuggets to everyone in the store. When asked where the nuggets came from, Mrs. Bowerman was more than happy to point to the mountain where her husband had his claim. Bowerman and his friend Dunn reportedly had been working their claim, quietly storing away any ore they may have found. The mine may have played out, or he may have found his gold elsewhere. Whatever the case, Bowerman was not shipping any gold.

Very soon, Mrs. Bowerman's idle chat set off a small gold rush. Within three days, 600 men were staking claims, looking for float, and digging holes along the mountain. Everyone was talking about Bowerman's strike and the very high-grade gold ore he had found, estimated at $70 thousand to the ton.

It was the summer of 1903, and the prospectors named their nascent town Bowerman. The town had a business district, a couple of hotels, about a dozen saloons and gambling halls, even a local newspaper.

Prospectors who descended on the area never found any high-grade ore. The low-grade ore they found was enough to keep them hopeful of making a strike. With all this activity, Bowerman still did not ship out any high-grade ore. His only discernable activity was to put a fence around his claim. He said he was afraid that anyone walking by could simply walk away with thousands of dollars in gold. Still, Bowerman did not ship out any ore. There was always an excuse.

By 1910 Bowerman was deserted. Today, only some crumbling buildings remain.

IF Bowerman did find a source of high-grade gold ore, he did little to exploit it, and it is still there, waiting to be claimed.

CRYSTAL
Gunnison County

Directions: At the outskirts of Marble, a small wooden sign points to a rough dirt road that heads east for eight miles to the site of Crystal. The first three miles are straight up a rocky road with no turnouts. Only one rough scenic road leads into town.

Prospectors first came to the area in the 1860's, but Crystal got its start in 1880 when prospectors formed a little silver mining camp in the isolated mountain meadow surrounded on all sides by extremely high mountain peaks: Treasury Peak, Mineral Point, Crystal Mountain, Bear Mountain, Sheep Mountain. When Crystal was a boomtown in the 1880's it was called Crystal City; its name was soon abbreviated to Crystal.

The Crystal River flows through the meadow. The famous Crystal Mill, built by G. C. Eaton, is on the far banks of the river. The mill is so picturesque that photos of it have been used for calendars, postcards, and it even graces the packages of a meat company's breakfast sausages. The mill supplied power for the Sheep Mountain Tunnel.

In the 1880's, a trail over Scofield Pass, and then on to Crested Butte, a distance of almost 20 rugged miles, was the only way to reach Crystal. This trail eventually became impassable. Today, the only road into Crystal winds through Marble and Carbondale.

The Gothic Miner printed the following sketch of Crystal on Saturday, 23 July 1881:

OUR MINES
SHEEP MOUNTAIN

Probably no camp in Colorado today can show the work or men employed for its age of development as Sheep Mountain. The hillsides fairly teem with men and the continued boom of the shots and blasts remind us of a battery practice. Below the mountain, where the new town of Crystal City has been laid out, we found a bustle and activity like that of the early day, and on Sunday last twenty-two cabins were under headway and more to follow. All of the boys turned out last week and worked the trail, so that traveling now from Scofield down the creek is fair. Holman & Morrison will supply meat to the camp, and Johnson Bros. will soon start a supply store. We have heard of no whiskey yet but it will soon "be dar." On Sheep Mountain last week by actual count there were 54 actual workings in progress, and by careful count over 500 men employed. The Grande Mesa & Rock Creek Co., represented by Mr. James, are working 35 men and preparing to do some extensive work. The tunnel to cut the Mammoth, Truckee, Dalton, and Cleopatra lodes will be worked from one side of the mountain and the shaft pushed from the top at the same time. They have a $500,000 paid in working capital, and the company is a strong corporation.

The Gothic Miner
Saturday, 23 July 1881

Crystal had at least 500 residents for 15 years. There were saloons, stores, an elegant hotel, pool halls, a school, and a local club, called the Crystal Club. The post office kept Fred Johnson busy carrying mail to and from Crested Butte in the summer by horse and the winter by snowshoes. Fred's brother, Al Johnson, ran a store, managed a hotel, and published the local newspaper, *The Crystal River Current*. The weekly newspaper was *The Silver Lance*.

In 1882 *The Colorado Business Directory* listed the following on Crystal:

Chrystal City [sic]
New mining camp in Gunnison County, 2 ½ miles below Scofield.

Burkley Mining Company, Charles Shuey, supt.
Cherry Smelting Company, project`d.
Harris & Company, gen`l mdse, and miners supplies.
Johnson Bros. store and saloon.
Scotland, Rev., assayer.
Snyder, H. C. & Co., real estate and mines.
Stone Mining Co., Geo. A. Stone, manager.

Crystal had some good mines. In 1890 the Black Queen was prospering. The ore was transported out on 150 sure-footed jackasses who trekked the 20 miles over Scofield Pass and then on to Crested Butte. The Black Queen's silver ore was displayed at the Chicago World's Fair in 1893. The Black Eagle, the Catalpa, the Farley, the Harrison, the Inez, and the Sheep

Mountain Tunnel mines were all good producers. The Lead Queen shipped ore until 1913, longer than any of the other mines.

In the boom years, the town's greatest problem was its inaccessibility. Winter snowstorms often cut off the town. The only roads into town would be covered in snow from 10 feet to 50 feet deep.

Crystal produced mostly silver, but small amounts of lead, gold, and zinc were also mined. Crystal was one of the towns that was destroyed by the silver crash of 1893. Because of the sheer beauty of the area, many people tried to hang on, unsuccessfully.

In recent years, a few people have discovered that this is a wonderful place to camp, hike, and take photographs.

DORCHESTER
Gunnison County

Directions: Dorchester is located on Forest Road 742 about ten miles on the north side of Taylor Park Reservoir.

Dorchester was founded in the year 1900 when gold was discovered in the Italian Mountain District. There were many mines in the area: the Bull Domingo Mine was on Italian Mountain; north of the Bull Domingo was the Enterprise; south of the Bull Domingo was the Star; and to the east was the Pie Plant. The Forest Hill Mine was southeast.

When the word spread that gold was discovered here, prospectors poured into town. The population soon peaked at 1,000. The miners worked all through the winter, facing the constant hazard of avalanches and blizzards. When winter snows hit, snowshoes were the only means of transportation.

After World War I, the mines closed, and the town was deserted. The winters have been hard on the buildings.

The site is now used for camping.

FLORESTA (Ruby, Ruby-Anthracite)
Gunnison County

Directions: To get to Floresta, take an old wagon road north from Crested Butte. Keep driving and when the road forks, stay left and keep going. After ten miles you will find yourself at an intersection: one fork goes to Irwin, and the other to Kebler Plass. Take the left fork. At one time there was an old sawmill here. Take this dirt road for about three more miles, keeping left at the next fork. You will come to the top of the mountain, and the townsite of Floresta will be at the bottom and at the end of a mountain valley. July and August are the best times to drive to Floresta. Until the middle of July, you will find the road wet from the snowmelt in spots. A pile of gray rocks, a stone structure, and some cabin remains mark the town site of Floresta.

A second route to Floresta requires a four-wheel drive vehicle with good clearance. Take an old wagon road that leaves Gunnison heading north towards Crested Butte for exactly five miles. At this point it will be another three miles. Turn left, taking an old wagon road through the old town site of Baldwin and Mount Carbon and then you will be going over Ohio Pass, but before you go over the pass, look to the east. You will see some huge rock palisades that were never finished. It was for the train track that went to the town of Irwin. The train track was only halfway finished in 1882 when all of the mines in Irwin played out and the townspeople left. From this point in the road, drive another mile and a half and then you will be at the turnout that leads to Floresta.

Floresta, which means forest in Spanish, was a town that grew around a coal mine named Floresta. The town sprang up sometime before the year 1900. There are conflicting dates: one source says 1902, but another says that it was before 1900.

When the town first started, it was originally called Ruby and Ruby-Anthracite, and then changed to Floresta. Floresta was a supply and shipping stop for the area. The town peaked at 250 residents, and a Gunnison stagecoach made regular trips in and out of town. When winter was at its worst, between January and April, much of the town's population left. Because the town was at the bottom of the valley, the snow would fall for days, and the snow pack typically was 25 feet deep. The coldest temperatures in the country are often in Gunnison County. In the winter, for the miners that stayed in town it was business as usual. They would go out and cut wood, but with the snow pack 25 feet high, the miners would be cutting off the tops of the trees. It must have been quite a sight come spring and summer to see these pine trees with their tops sheared. To this day, you can still see trees with missing tops.

A large deposit of fine anthracite coal started at Floresta and stretched as far as Crested Butte and Somerset. The Colorado Fuel and Iron Company owned and operated the Floresta Coal Mine. They helped to develop the camp, and they built the railroad into town. The saddest casualty of this railroad construction was a little boy who just wanted to watch the railroad workers as they were laying the rails and building the trestles. He lived in the area and was fascinated by the railroad. When the workers were putting together the great trestle and the beams were being lifted into the air, the beams fell on the little boy, killing him. This was a great tragedy for such a small camp.

Floresta was a prosperous coal town with a bustling business district, and its cabins were scattered all over the gulch. Eventually the price of coal fell, and the railroad was closed down in 1920. The town of Floresta is so remote that without a railroad, it meant the beginning of the end of the town. By 1936, the town was deserted.

GOTHIC
Gunnison County

Directions: Gothic has an altitude of 9,500 feet. Leaving Crested Butte going north on State Highway 327, drive nine miles through the East River Valley. After nine miles, exit the highway, and turn onto a dirt road called Gothic Road. The remains of Gothic will be on the east side of the East River where it meets with Copper Creek.

The town of Gothic got its start in June of 1879. A prospector named James Jennings made a rich strike at the head of Copper Creek, four miles from Gothic. The mine, which he called the Sylvanite, held rich pockets of silver wire. It was literally silver wire; you could break the rock in half, and the silver wire that was laced through the rock would still hold everything together. The ore was worth $15 thousand a ton, until the mine played out in 1885.

The Sylvanite Mine was discovered in June of 1879, and within a week over 100 makeshift tents and cabins had sprung up. Gothic grew rapidly and soon had a business district lining its long main street. By the end of the summer, 170 buildings had been built. A telegraph line was run to Gothic the first of October, 1880, and Mr. Turrell, the town druggist, managed the telegraph office.

Like many small mountain towns, Gothic frequently was cut off by deep winter snows. During the winter of 1994-1995, for a recent measurement, nearly 54 feet of snow fell.

By the year 1881, more rich gold and silver strikes had been made, and the population exploded to 1,000. Prospectors continued to stream into the town by the hundreds. Gothic was now the top producing town in Gunnison County. Every type of frontier business was started on Main Street: saloons, gambling houses, stores, hotels, mills, stables, schools, etc. When night came, the men would ignite a huge bonfire in the middle of Main Street and socialize. The population soon peaked at 8,000, with basically no crime except when one man was murdered, fighting over a lot, and one man was lynched, who was probably the man who got in a fight and shot the man over a lot.

The town had a newspaper, *The Elk Mountain Bonanza*, which was renamed *The Gothic Miner* in the spring of 1881.

The Gothic Miner printed the following on Saturday, 28 May 1881;

GOTHIC'S FUTURE

Gothic has fairly started now into its summer's work, and daily everything goes to strengthen it and make its existence a certainty beyond a shadow of doubt. In the past no puffing has been attempted; the truth is good enough, and now only a fair showing of what is doing will be sufficient evidence that we have one of the strongest camps outside of Leadville. The faces we see daily returning after a winters absence, are those of men who have been here for the past one, two or more years. There is well-founded faith in the prosperity of Gothic, which brings them here as a load-stone draws the magnet; and no excitement, however great, can attract them to other fields. Improvements are going on, new houses rapidly building, streets being cleared and made passable, miners are out daily, and contractors are beginning to look for men to work their properties. We assert that at least 1,500 men will be employed by others in mining alone this summer; that 500 men will be prospecting and working for themselves, and that $200,000 at the lowest possible estimate, will be expended in developing mines. These estimates do not include the vast amount of work on the smelters, particularly Avery's smelter; nor does it include the men or money required to carry on the large business establishments of the town. All of the surrounding camps are tributary to Gothic and will receive of them revenue from their work. This hastily drawn sketch of Gothic's future does hardly touch the most superficial points of its certain wealth.

Shideler & Morrow, a handsome office building in rear of blacksmith shop on Main street, 14 x 24, to be used as a mining exchange.

Whitman & Moak, a new face and siding to the building lately occupied by Messrs. Dudly & Steel. It makes a handsome appearance and it is intended to be used as a first-class restaurant.

W. Hobbs, building on Main street next to the corner of Third, a large hall. Its use is not yet decided, but it will be one of the largest buildings in town.

Barthel & Burchart. Mr. Barthel is erecting to the rear of his dwelling on Main street, an addition, two stories 16 x 24. In the rear of this building Barthel & Buchart are building a large and substantial bakery.

Gothic hotel, a large extension on the hotel. The lower part will be used for bath rooms and barber shop, and the upper story for sleeping departments, which will increase the capacity of the present commodious house.

M. A. Buckles, on upper Main street, an 18 x 32, two-story frame. It is substantial, and is intended to be used as a lodging house.

Calwell & Blewitt, 80 foot bolling alley [bowling], in rear of saloon.

Prof. J. P. Ralston, putting in glass front in the store of Fletcher Bruce, on Main street.

Brown & Co., a large handsome two-story store-house on Main street between Smelter and Third streets. To be used as a meat market.

Beardsley & McNeal, a large substantial blacksmith on the corner. [?]

The Gothic Miner
Saturday, 28 May 1881

So much was going on that the town soon became a shipping and supply center for the other little camps in the area. President Ulysses S. Grant spent the night in Gothic in 1880 during his tour of the Western boomtowns. He was greeted by a parade of cheers and salutes of gunfire. The President stayed at The Olds Hotel. The owners of the hotel were so excited that the president staying there, that they renamed the hotel The Grant. Later, their excitement waned, and they changed the name to The Gothic Hotel.

The Gothic Miner printed the following on Saturday, 2 July 1881:

GOTHIC'S FOURTH

Through the energy of the proprietors of the Half-Way House Gothic is not behind its sister towns in celebrating the Fourth of July. Turn out boys, and have fun. You know where the Half-Way House is. A good foot bridge passes over East river and there will be no excuse for not enjoying yourself. Scofield and Elko will also send portions of their inhabitants and a crowd will attend.

The exercises will begin at 11 o'clock sharp.

Reading the Declaration of Independence by C. H. Garstin.

Introduction of the orator of the day by Mayor J. G. Pease.

Oration, by ex-Governor Thomas Carney, of Kansas.

Dinner.

After the heat of the day the races and amusements will begin.

1st- Sweepstakes, $1.00 entry

2nd-Fast burro race, .50c entry.

3rd-Slow burro race, .50c entry.

Rifle and pistol shooting.

1st match-Rifles, 200 yards, one shot, 50c entry.

2nd match-Pistols, 30 yards, one shot, 50c entry.

3rd match- Rifles, 300 yards, one shot, 50c entry.

Jumping, pitching quoits, foot races, and towards the evening fireworks.

The above programme shows the manner in which the day will pass, and in the opening with the gentlemen whose names are given a good treat is promised to all. The oration of Governor Carney should be heard.

Stark & Glick are crowded with new goods. In addition to their groceries on Thursday they received a complete stock of boots and shoes.

Messrs. Wait and Patterson, the contractors for making the Crested Butte and Gothic toll road, have very successfully completed their work. It is now the best road in the county.

Barlow & Sanderson's fine stage line with its splendidly kept stock, came dashing into town on Friday last week, and yesterday took charge of the mail. Its trips are daily.

Billie and Johnie are crowded nightly at their keno tables. In fact it is a pleasant, quiet and amusing game. You can hear the --- call of "first ball" from the Office saloon.

Slop and refuse barrels are accumulating in the rears of the hotels and restaurants to the delight of the millions of flies collected. When one is filled, instead of being removed, another is added to be filled.

The Cochetopa Placer and Gold Quartz Mining Company, consisting principally of Messrs, G. E. Sprout, J. H. Adams, Earnest Schluter, Dr. O.R. Stoddard and T. J. May, are developing their number of placer claims in the Cochetopa range.

Mr. J. S. Adler, the well known jeweler and watchmaker, has transferred his establishment to Stark & Glick's.

The Gothic Miner
Saturday, 2 July 1881

The Gothic Miner printed the following social news on Saturday, 23 July 1881:

LOCAL ITEMS
Billie the "Kid" is dead.
You, Brad! Look out dar.
John it's time to get back to camp.
McClancy has got a herd of burros now.
Atwood has gone down Rock creek fishing.
Rumor says Cooper has gone to get married.
Mansfield you got a pretty good trade on the watch.
Luona stock jumped from 15 cents to 50 cents this week.
Shove's buying Luona stock, Banning's bearing it, Moore's trying to get even.
The new sidewalk in front of the Bon Ton catches the eye of pedestrians.
Fay Hoyt caught sixty trout Thursday. Guess he holds the best hand so far.
D. & R. G. stock quoted 105 ¼ at New York. Pretty fair stock and a pretty fair road. Eh?
It puts us in mind of old times to see the guests of our first class hotels sleeping on parlor floors.
Moore & Jones, our saw-mill men, have moved their mill up to the Half-Way House near a body of good timber.
The church is looking like a church.

The Gothic Miner
Saturday, 23 July 1881

In 1882, *The Colorado Shipping Guide* listed the population at 800. In 1884-1885, only a few years later, all of the area mines went dry. All of the townspeople left except for Garwood N. Judd. He came to town in 1880. In the early years of Gothic, he was a model citizen who wanted to be mayor. So did Lew Wait. Everyone agreed that it would be fair if they rolled dice and the high number would be mayor. Lew Wait won the roll of the dice and the office of mayor. In 1885 when everyone else left town, Garwood H. Judd stayed on and appointed himself mayor. He eventually had to move to Crested Butte when he became too old to take care of himself, but he left instructions that after his death in May 15, 1930 that his ashes be scattered in town.

During the mid 1900's, the Rocky Mountain Biological Laboratory constructed a few buildings to study the wildlife in the area. They tore down most of the abandoned buildings of Gothic.

This is still a perfect place to camp in the summer. Gothic Mountain towers over the town site at an elevation of 12,570 feet. If you want to see some beautiful scenery, you should try this wild four-wheel drive road from the Crystal River Valley going over the old Scofield Pass and on to Gothic. Scofield Pass was closed, hopefully just temporarily, in the fall of 2000.

HILLERTON
Gunnison County

Directions: Hillerton is 16 miles north of Pitkin, and two miles north of Tin Cup on Forest Route 765.

In 1879 a few seemingly good strikes were made: The Little Earl, The Adeline, the What Is It. Prospectors and other fortune hunters rushed to the area, and a town formed. The town was called Hillerton after Edward Hiller who helped found the town. He was a banker who had built a bank in town. Within a few months, three stores were built. One was run by Mr. S. A. Sutton. There was a nice hotel called the New England Hotel, which was owned and run by Mrs. M.J. Small, three sawmills, and many other businesses. A toll road offered a shortcut to Ruby and Crested Butte. The population peaked at 1,500 by 1880.

The Hillerton Occident newspaper was published by Henry Olney in June of 1879. Tin Cup, then called Virginia City, was two miles south of Hillerton. Tin Cup was started a month after Hillerton, but its mines were bigger and more productive. Henry Olney packed up his printing press and moved to Tin Cup within a few months.

In 1882 S.J. Klauber ran the post office and the liquor store. Mr. Sutton's store was the only remaining store. The hotel was still open, and the Virginia Smelter was still working. The 1882 *Colorado State Business Directory* reported a population of 25.

After a few years, people started to pack up and go to Tin Cup. The mines around Hillerton were too underdeveloped, and the ore was not worth mining. Some of the residents dismantled their log cabins and dragged their logs with them. By just looking at the site, nothing is left to the plain eye, so you need to look for more subtle signs. Reportedly, many of the cabin roofs were made of sod and when the actual buildings were gone, the thick sod would settle into a softly rounded mound, noticeably raising the site of where the house once stood. The sod enriched the soil and made the topsoil deeper, so that sagebrush and plants would grow much taller than anywhere else. Also, look for broken glass and pottery. This helps you find the actual sites.

Abbeyville was another camp that grew up between Hillerton and Tin Cup at the same time these two camps began. The C.F. Abbey Smelter was built and in operation in 1881, and around this Abbeyville was formed.

Abbeyville died faster than Hillerton, leaving little information about its life and residents.

The Carbonate Weekly Chronicle printed the following on Hillerton 13 March 1880.

HILLERTON is situated two miles below Virginia City, in a beautiful, level valley called Taylor Park. The scenery about Hillerton is picturesquely beautiful, and two fine streams run, one on each side of the town. A large smelter is already erected, and the town has a commodious hotel, a bank and large stores. Hillerton is destined to become a much sought-after place as one of the summer resorts of Colorado. Near the town rich gold placers have been discovered, and Messrs. Pease & Hall have built a ten mile flume to work them. Facing the town stands the majestic Amazon Hill, abounding in mineral, but few claims have yet been developed. As of rivalry between the two towns, I do not consider there will ever be any, as there is ample place and enough mines in the district to insure the prosperity and success of both places. As to the time of starting for the Gunnison, for the information of newcomers I would advise not to venture on the trip at present. Yesterday four young men passed through here for Ruby Camp and the Ute Reservation. They had jacks loaded with provisions and were ready for the campaign. It is no use to leave Leadville for this region before the middle of April, and under no consideration before the first. At the present time there is no way of reaching this place with jacks or teams inside the distance of two hundred miles. By the middle of April the trips can be made in half that distance over good roads. The Alpine and Virginia City Toll Road Company guarantee to keep the road open after April 1. The Virginia, Hillerton and Roaring Forks Toll Road Company, incorporated with a capital stock of $50,000, will open the road to Roaring Forks as soon as the weather will permit. The toll road company from Hillerton to Jack's Cabin, which is the shortest route to Crested Butte, Gothic and Ruby City, will soon be in condition. The trip should by all means be deferred until the middle of April. As to the remarks made by some parties that if they wait until late, all the claims will be taken, let me tell you that when they leave the McKenzie valley, you are traveling for a hundred miles over as good a mineral country as can be found, and if a hundred thousand men were to come to Gunnison in the spring, they would not more than half prospect the mountains. So there is no fear of not finding a chance to discover mines.

The Carbonate Weekly Chronicle
13 March 1880

IRWIN
Gunnison County

Directions: Leaving Crested Butte on the Kebler Pass highway, go west for about ten miles. There will be a dirt road on your right, it may be called the Lake Irwin Loop. Take this road and stay to your right. After a mile you are at the site of Irwin.

Dick Irwin and some of his prospecting friends discovered a lode of rich silver ore in the fall of 1879. They sent their ore to Alamosa, the closest railroad stop, and from there it went to Denver to be assayed. About a month after the ore was assayed, hundreds of men showed up from all over and started staking claims, and even in the snow they started chopping down trees and building cabins. The snow pack was 90 feet deep when the men built their cabins. When spring finally came and the snow melted, the tree stumps that the men had thought that they cut to the ground were ten feet tall.

At the same time that Irwin and his friends had come to the area, in 1879, a prospector named Fisher had come to the area with his wagon and team of oxen. He discovered some rich ore deposits, and just after being in the area a couple of days he staked two claims: the Ruby Chief and the Forest Queen. By July 1879 the Ruby Mining District had formed. By fall when the weather started to get cold everyone in the area left for the winter except 50 people.

With the onset of winter, the snow came down, burying the cabins. Fifty feet of snow came down before it stopped. The miners had to dig tunnels from their doors and chimneys to the surface. In the early spring groups of prospectors, impatiently hurried to Irwin to stake their claims. In one story, newly-arrived prospectors made it into the snowbound camp, but they saw no buildings. They saw a man standing next to a tunnel in the snow, and they saw scattered holes in the snow. Some had smoke coming out of them. The new prospectors asked the man if he knew where the town of Irwin was. The man told the prospectors that they were in Irwin. When the prospectors asked where the post office was, the man replied, with a chuckle, "It's the next hole to the right."

By the spring of 1880, hundreds prospectors had come to Irwin. Tents and cabins were scattered all over, and a main street was platted, dotted with stores, saloons, and businesses.

Next to Irwin, less than a quarter of a mile away, a businessman from Leadville said that he was going to form a town site, calling it Ruby City. He had promised that Ruby City would be big, a great investment. He would himself build a store, a theater, an office building, and a six-story hotel. He sold as many property lots as he could and then skipped town. Next to Ruby City, the towns of Haverly and Silver Gate went up. Within a year, all of these towns grew into one. Some people called it Ruby because the type of silver found in the area was ruby silver. In the end, the one big city was called Irwin.

Some good mines were established near town: the Bullion King, the Last Chance, the Lead King, and the Monte Carlo. Across Brennard Lake a little farther from town were the Ruby King and the Ruby Chief. The Forest Queen was one of the best mines in the area. Someone had offered the owner $1 million for it, but the owner refused. In 1932, the mine was sold for $40.45 in back taxes.

Wild Bill Hickock and Theodore Roosevelt visited town, creating much excitement. General Grant came to visit town. The local newspaper, called *The Pilot*, was established in June 17, 1880. There were three churches: an Episcopal, Presbyterian, and Methodist. The town boasted big metal street signs and even fire hydrants as protection against the ever-present fire danger that threatened so many mountain towns.

The boom years here were 1881 and 1882. In 1881, Main Street extended the length of the gulch for at least a mile. Attractive, solidly built buildings graced Main Street. There were 23 saloons, some with mahogany bars and tables and with mirrors throughout.

The Colorado Business Directory for 1882 listed the population at 500.

In 1883 people started to leave town. By 1884, the sporadic mining in the area was not enough to support a city.

By 1896, the population of Irwin had slid to only 200, when *The Colorado Business Directory* for 1896 reported the following:

IRWIN

Mining town in Gunnison county, one-half mile from Ruby, Anthracite branch D. & R. G. R. R. Population 200. Daily mails. Altitude 10,250.

City Water Works Co, water supply, F W Fuller, mgr
Fuller, F W, mgr Forest Queen Mining Co
Hinkley, W S, postmaster
Irwin Milling & Power Co, ore sampling and concentrating works
Kelley, R R, hotel
Ropell, P F, mgr Mt Gunnison mine
Whipp & Samsel, leasers Ruby King mine

Today, nothing remains of Irwin.

MARBLE
Gunnison County

Directions: From Carbondale on Colorado Route 82, take Colorado Route 133 to Redstone, which is south 17 miles. At Redstone, drive five miles past the town. On Gunnison County Road 3, you turn east and drive for five miles to Marble. To reach the quarry, just drive three miles farther and you should be there.

In the late 1870's, two small mining camps by the name of Yule Creek and Clarence had formed here, mining both gold and silver. The prospectors had discovered a beautiful bone white marble in 1873, but did not yet quite know what to do with it.

By the early 1880's the two camps grew into one, and the merged town was christened Marble. The town newspaper, *The Marble Times,* was first published in 1892.

Silver and gold production ceased' and the miners started to take out the marble and sell it. They hauled the marble by wagon in the summer and by sled in the winter.

In 1890, the Kelley Brothers opened the first marble quarry. Then Mitchel and Fine opened the second quarry. J.C. Osgood opened the third quarry. Marble from these White House Mountain quarries has been used all over the country. In 1912 the Yule Marble quarry reached its peak, employing 2,000 people. Marble from this quarry was used for the Tomb of the Unknown Soldier, the Lincoln Memorial, the Federal Reserve Bank in Denver, as well as the Denver City and County building.

The town of Marble had many misfortunes.

In 1925, a fire destroyed the mill. Demand for the town's marble declined. In August of 1941 Yule Creek flooded, and the creek washed away part of the town. The disaster forced the quarries to close. The population was constantly changing until there was basically no one.

In 1990 some Denver men reopened the marble quarries and were soon shipping out 5 x 5 x 8 foot marble blocks weighing 17 tons each. They have been shipping out about a thousand of these blocks a year.

Every summer an art students group from Denver conducts workshops and a marble-sculpting exhibit. It lasts for a couple of weeks.

Collectors can find excellent, large blocks of marble along the road and creek bed, chunks that have slid off wagons and trucks over the past century.

Remains of the old town resemble Roman ruins. The buildings that had been the original town site had been made of marble and from numerous mudslides, rockslides, avalanches, and floods. Much of the old town is just a few pieces here and there of walls and foundations. Around the old town are some occupied homes.

NORTH STAR (Lakes Camp)
Gunnison County

Directions: Coming from the south of White Pine, you take a four-wheel drive trail heading northeast for a little over a mile, and you will reach North Star. The road is very narrow, and you will need a vehicle with clearance. The road runs the length of Galena Gulch with an old two-foot rock wall built up, which is called cribbing, on both sides for most of the way. The road is so narrow that you will be unable to turn around. You have to cross a wooden bridge – it should be safe, but you should check it out before going on it.

In 1878 prospectors discovered rich galena ore, and then in June 1879 the North Star Mine was excavated. The little mining camp that had formed here had been called Lakes Camp, but after the North Star Mine began producing and the camp started to grow into a town, the people here called it North Star. The town of North Star was built on property owned by the May-Mazeppa Mine, which also owned much of White Pine and a large ranch, with stables, bunkhouses, and outbuildings that held mining machinery. For area news and gossip, the residents of North Star read *The White Pine Cone*. For entertainment, there was the Soup Bone Musical Club. Members would sing, play the harp, the guitar, and a few other instruments. The 1880's were prosperous and booming, but in 1893 the Silver Crash caused all the mines to close, and the people left.

In 1901, the mines opened again, and men worked around the clock. The mines were open on and off until eventually they finally closed. When you first come into town, the first building on your left will be the Leadville House. It had a tall white false front. Next to it was the post office. It was built with logs. Across the street were a couple of cabins. The main street that went through town was called Main Street.

North Star was last known to be on private property, so if the road is blocked off you may have to ask permission to get in. Last known, the May-Mazeppa property must be passed to get to North Star. The site is as pretty as its name.

Today, the legend of Snowblind Gulch recounts the lost gold of two prospectors who stayed too late to escape the fury of an approaching mountain winter. Legends say that in 1860 when the first prospectors wandered through North Star and Tomichi, two prospectors stayed for the summer. They each panned out a pound of pure gold a day. When the cold weather set in and winter was upon them, they had to leave but they had stayed too long. No one knows if the snowstorms or the Indians killed them. Twenty years later, when people settled in the area, an old sluice box and mining equipment were found next to the stream.

The White Pine Cone printed the following article on 22 February 1884:

Thirty-five feet of snow in the Elk Mountains and four months of winter yet to hear from. The boys over in that vicinity should come to White Pine, the best winter camp in the state.

A great quantity of ore is daily taken out of the North Star, but the recent storms have rendered the roads temporarily impassible. Unless it should storm again shipments will be resumed next week.

The White Pine Cone
22 February 1884

The White Pine Cone printed the following on 11 April 1884;

That was a narrow escape for John Turner and Watt Clasby, last Saturday, while at work on the Silver Trowel. In lieu of a shaft house the boys had a large tent placed over the shaft. In the tent they had a large heating stove—but they don't have it any more. Before going down the shaft to work, Watt Clasby placed two sticks of giant powder on the stove to thaw out. They had been at work but a short time when they received the information that the powder had thawed! A loud explosion was heard followed by a deluge of snow and debris upon the men below. As soon as they could reach the surface the cause of the racket was apparent. The powder had exploded, blowing the stove and a portion of the tent into the "immensity of space." Some of the miners working near, heard the explosion and saw the tent suddenly collapse, and hurried to the Trowel, expecting to gaze upon the mangled corpses of Messrs. Turner and Clasby. The accident might have proved fatal, and the boys are to be congratulated upon escaping so fortunately.

The White Pine Cone
11 April 1884

Waves of scarlet fever, typhoid, small pox, and other diseases swept through the mountain towns, sometimes killing all the children of struggling families.

In the sad death of little Georgie Hutchison scarlet fever has numbered another victim. The dread disease has carried off a great many children during the past month in this state, and the greatest care should be taken by our people, particularly parents, to prevent its spread. In Gunnison, the authorities resorted to strict quarantine, and the same care should be exercised here. Scarlet fever is as fatal as small pox, and is equally as contagious, and no one who comes in contact with the disease should under any circumstances be allowed to enter a house where there are children until the person's clothes are thoroughly disinfected. By such precautionary measures only can the spread of the disease be prevented.

The White Pine Cone
11 April 1884

OHIO CITY   (Eagle City, Gold Creek)
Gunnison County

Directions: To reach Ohio City from Gunnison, take State Highway 50 head east about ten miles until you reach Parlin. From Parlin, take County Road 76 northeast for almost 11 miles until you reach Ohio City. It will be right past Forest Route 771 on the left side of County Road 76, maybe a few hundred yards on an unmarked road. Or if you are coming from Pitkin, take County Road 76 southwest for six miles. Ohio City will be on the right side of the road a few hundred yards down an unmarked road. Today, Ohio City is more a ghost town that a lost city.

Ohio City was founded in 1879 when prospectors discovered silver deposits. In the 1860's placer gold had been found in the area, but there was not enough to trigger a rush or a town. Eagle City had been the original name for a short time, and then Gold Creek briefly, and finally Ohio City. By 1880 Ohio City had turned into a boomtown. There were at least 50 log cabins and tents scattered around the town. There was a hack-run stage. The D & R G Railroad made daily trips from the Alpine Tunnel to Pitkin and then to Ohio City. From Ohio City, the train went to Parlin five miles away and on to Gunnison.

By 1882 the *Colorado State Business Directory* reported a population of 150 and listed the following businesses:

Atchison, Sponsolier and Whaley, mining brokers
Dodge and Company, general merchandise
Flick and Strand, produce
Gooden and Waterman, hotel
Hess, Jacob Postmaster
Legal Tender Stamp Mill
Ohio Mining Company, Colonel Sharp, superintendent
Sultana Milling Company, L.B. Martin, superintendent

Ohio City was prosperous and faced an optimistic future until the Silver Crash of 1893 closed down the mines.

In 1896 gold was discovered in Ohio City, igniting another boom. For this boom the town had a sawmill, so all of the new buildings and homes were made of sawn lumber, unlike the log structures that marked the 1880 boom. The Roller, Calumet, and Eagle mines were right next to the city. A little way up Gold Creek was the Carter Group of mines, which had 79 mining claims. It was called the Gold Brick district. Out of its mile and a half long tunnel, a $3,000 gold brick was shipped out every two weeks. The company spent $250 million to develop its mines. It had 20 stamp mills. These mines ran a profit for the next 20 years. A mile farther than Carter on Gold Creek was the Raymond Mine. By 1916 the Raymond produced over $7 million in gold. Another mile farther on Gold Creek was the Gold Links Mining Company. It had a 40-stamp mill, the biggest mill in the district, with many scattered buildings worked by 200 men. The property spanned over 6,000 acres. A mine tunnel, which started at the Creed, extended over 4,000 feet. This mine produced about a $1 million in gold and silver. It has been abandoned for a long time.

As of July 2001, only 86 people lived in Ohio City, and there was only one store, The Ohio City Store. Its post office, in many ways the center of the community, closed 31 July 2001.

By going through the small town, you might not see the historical significance of days gone by. Look a little closer.

SCOFIELD
Gunnison County

Directions: From Crystal, take Forest Road 317 for four miles following the South Fork River. It is located at the head of a creek by Crystal River. It is 16 miles north of Crested Butte.

Prospectors discovered silver here in 1872, but because of the decline in silver prices in 1873, prospectors' fear of Ute Indians, and because the area was hard to get to, nothing was done to develop it until 1879. In 1879, Chief Ouray assured the white men that there would be no more Indian attacks, and then more silver was found. In August 1879, B.F. Scofield led a group of prospectors in forming the town of Scofield. It was located between Crystal City and Elko. When the town was formed, it already had 300 residents.

Elko was a camp just a few miles from Scofield and five miles from Gothic. Only a small group of prospectors lived here, and it had only two cabins. But Elko lasted about five years longer than Scofield.

In 1880 General Grant and the former Colorado Governor Routte came for a tour of Scofield. There was not much to see beyond a dark, deep, and foreboding canyon on the north side of town, nicknamed "Sonofabitch Basin." So, the town brought out a barrel of its best whiskey, and everyone just had a good time.

In 1882 Scofield had 150 residents according to the *Colorado State Business Directory*. For a town that was so isolated, it still had a lot of businesses: a post office, hotel, general store, restaurant, barber shop, blacksmith, and carpenter. Daily stages traveled to Crested Butte.

Scofield was at the base of Scofield Pass, and the town was a rest stop for travelers before going over the pass. It was a treacherous, scary pass. In winter the snow pack was 40 feet deep. Transportation was a problem, and when the ore in the mines turned low grade, the town's population started to dwindle.

By 1885, the town was deserted. Little remains of the town.

Winter is rough in this area; avalanches are common and run 200 feet deep. Midsummer -- July and August -- is the best time to visit Scofield.

The White Pine Cone reprinted the following article about a winter snowstorm on February 22, 1884;

This winter threatens to lay over the terrible winter of 1879-'80 for deep snow, and perhaps to excel the winter of 1881, when the snow was nine feet deep in East river and five feet deep in the Gunnison valleys.- Over thirty-five feet of snow has fallen as Elko. The roads to all mountain camps are blockaded, the snowshoe express and telephone being the only communication. No mail has gone to Scofield for a week, it being too dangerous to travel the trail on account of snowslides. No mail has arrived from Crested Butte for three days. Running a newspaper under such circumstances is not one of the pleasures of life. — *Gothic Record.*

The White Pine Cone
22 February 1884

BURROWS PARK
Hinsdale County

Directions: The town of Burrows Park is a quarter of a mile below Whitecross on County Road 4.

At an elevation of 10,700 feet, the town of Burrows Park was located in a high alpine meadow, which was called Burrows Park. Formed in 1877, the town was named for Charles Burrows, a prospector who had come to the area in 1873. The town of Whitecross was a quarter of a mile to the north. In the winter, the town's population disappeared until spring. The population peaked at 200. The Oneida, the Napoleon, and the Undine were the area's silver and copper mines.

In 1882, *The Colorado State Directory* reported only two businesses in its listing for Burrows Park:

BURROWS PARK
A mining camp in Hinsdale County, 7 miles east of Animas Forks and 22 southwest of Lake City. Tri-weekly mails.

Williams, Mrs., hotel.
Ellsburg, P. N., gen'l mdse.

After the silver crash of 1893, everyone left. The town has been reduced to rubble except for two cabins the forest service has tried to save. The setting of this town site is strikingly beautiful.

CAPITOL CITY (Galena City)
Hinsdale County

Directions: From Lake City, take County Road 20 west for nine miles along Henson Creek, until you reach Capitol City. The road may be a little winding, narrow, and rocky, but any car can make it.

In 1877, prospectors discovered rich veins of silver, and a mining town was formed called Galena City. The town site consisted of 200 acres of lush meadow bordered by towering mountains, at an altitude of 9,480 feet Prospectors came by the dozens to seek their fortune after hearing about the rich strikes here. The residents had high hopes for their town. George Lee and some of the other town residents were very ambitious. They decided to change the name of their community from Galena City to Capitol City, a public relations move to challenge Denver as the future capitol of Colorado.

The town grew rapidly, and in May 1877 a post office was established. George Lee built a sawmill to support the town's growth. Soon, stores, saloons, a few hotels, restaurants, and smelters were built. In 1881 telephone service was hooked up for the town. By 1883, a school was built at a cost of $1,500. The population peaked at between 700 and 800, and the town had been overbuilt; there was enough room in town to hold at least a thousand more people. They had built the town, and they were waiting for the people to come. Daily stages ran to and from Lake City, nine miles away for $1.50. It cost $1.75 for the stage trip to Animas Forks, only 12 miles away.

George Lee had built a big, beautiful brick house, which everyone in Capitol City called the Lee Mansion. Lee himself called it his "Governor's Mansion," a hint of his political aspirations. The Victorian mansion had a large bay window in the front living room, a small theater with its own orchestra pit, and a ballroom. The bedrooms were on the second floor. The stable and many outbuildings were built of the same fine red brick as the house. The brick had been shipped from Pueblo at the extravagant price of a dollar a brick. The Lees held fancy parties. George Lee clung to the dream that Capitol City would become the state capitol of Colorado, and his home would some day be the governor's mansion.

The mines in the area had been producing well. Two mines, the Capitol City and the Yellow Medicine, each produced a few hundred thousand dollars of good quality ore. Other mines included the Great Eastern, the

High Muck-A-Muck, the Incas, the Morning Star Lode, the Ocean Wave, the Polar Star, and the San Bruno.

When silver was devalued in 1893, the town was devastated and almost everyone left except the Lees. They refused to give up their dreams and stayed in the dying town. By 1896, only 50 people were living in Capitol City.

The Colorado State Business Directory for 1896 listed the following for Capitol City:

CAPITOL CITY
Old mining town in Hinsdale county, 9 miles from Lake City, the nearest railroad point and banking town. Stage six times per week. Population 50.

Brunner, Mrs Lydia C, postmistress
Guionneau, Benj, assayer
Guionneau, Bros, gen mdse & lumber

In 1900 gold and copper were discovered, breathing new life into Capitol City. Old mines were reworked, and new ones were opened. During this second boom, the Moro and the Ajax were the best producers. This continued for almost twenty years, and then the mines stopped producing. Everyone left town, and this time it was for good.

In October 1920, the post office closed its doors, and the postmaster packed his bags and left.

Almost nothing is left of Capitol City. If you look in the underbrush you may find some sign of a foundation, or you might see small shards of glass on the ground.

CARSON
Hinsdale County

Directions: Coming from Lake City, take County Road 30 southwest for about 11 miles until you reach Wager Gulch. Any car will do until you reach this point. With a good four-wheel drive follow Wager Gulch Trail south on County Road 9. After a half-mile or so, it turns into Forest Route 568. Follow the Wager Gulch Trail / Road south for about 3 ½ miles south, up the worst, narrowest jeep trail that you have ever been on, to the summit near Bent Peak. When the trail starts to get rough, walk the rest of the way.

In the best of weather, Carson was nearly inaccessible, but winter storms would completely close down the town. It was at an elevation of 12,000 feet. The only way to Carson was over a rough steep road from the other side of the Continental Divide. Carson is still a very difficult mining town to reach.

In 1881 Christopher Carson made a rich silver strike on the top of the Continental Divide. His claim, the Bonanza King, was built on an iron dyke and attracted deadly electrical storms on summer afternoons. The town was called Carson Camp at first, until it grew, and then it was just called Carson. It had a hotel and a general store.

The wagon trail up Wager Gulch was rough and steep. In 1887 the Lost Trail Creek Road was built.

The Bonanza King was a good producer, but the St. Jacobs was better. By 1892, the St. Jacobs had produced $300,000 in ore, and in 1898 it had produced $190,000 in ore during that year alone. Other mines were the Bachelor, the Chandler, the Cresco, the Dunderberg, the George III, the Iron Mask, the Kit Carson, the Legal Tender, the Lost Trail, the Maid of Carson, the Thor, and the St. Johns.

Carson was built on silver mining. In 1893, the silver crash almost destroyed Carson. The town emptied out, and nothing was heard until 1896, when some good gold strikes were made. Men returned to the area to work the new mines and rework some of the old abandoned ones. The town's general store and saloon shared one building. A boardinghouse and post office were in another. Another large building was the Bonanza King's office. At least 150 mining claims were being worked. Mines were scattered all over, along with miners' cabins.

During the 1896 silver boom, the miners wisely built their cabins and town buildings at a lower level on Wager Gulch Trail, at an elevation of 11,500 feet, minimizing some of the problems that plagued the first town site. The earlier, almost inaccessible Carson was on top of the summit at an elevation of 12,000 feet. Winter paralyzed early Carson, and lightning was a deadly threat throughout the summer. The new Carson had around 500 residents.

By 1903 the Carson mines had been played out, and the town emptied out.

The old town of Carson at the summit has crumbled. The newer Carson, below the original town, has a few buildings still standing. The Lost Trail Creek Road from the south is too overgrown to use. To visit Carson, you must use the Wager Gulch Trail.

The very inaccessibility of the town that was such a problem early in its life now has helped to protect it.

HENSON
Hinsdale County

Directions: To reach Henson, take County Road 20 west, up Henson Gulch from Lake City for about three miles.

In 1848 the U.S. Army sent out an exploration party led by Col. John C. Fremont. The party reported seeing gold in the Lake City and Henson area. Decades later, in 1871, Charles Goodwin, Joel K. Mullen, Albert Mead, and Henry Henson made a rich gold strike, naming it the Ute-Ulay Mine. Prospectors rushed to the area, outraging the Ute Indians. The area was Indian Territory, so all the white settlers left the area temporarily until 1873 when the Brunot Agreement was signed. This agreement opened the area to white settlers. The founders of the Ute-Ulay Mine returned to work their claims. Many more prospectors followed, and just on the other side of the Ute-Ulay, the Hidden Treasure Mine was discovered.

In 1876 the founders of the Ute-Ulay Mine sold it to the Crooke Bros. in Lake City for $125,000. The Crooke Bros. then built a lead smelter to process the ore from the Ute-Ulay and the Polar Star mines. By 1880 the Ute-Ulay had been so successful that its owners were able to sell it for $1,200,000.

In 1877 the Uncompahgre Toll Road and the Henson Creek Road were finished. The road was so successful that the area settlers contributed funding and manual labor to extend the road over the mountain range, reaching the Animas Forks Road and linking that to Ouray and then to Silverton. Today, this historic byway is a 65-mile four-wheel drive road called the Alpine Loop. It traverses the Continental Divide at an elevation of 12,000 feet.

In 1880, the town was platted and named Henson after founder Judge Henry Henson. The town of Henson was nestled in a very narrow gulch. There was very little room to spread out in this narrow gulch, and the town grew very congested.

Henson was a tough mining town. Fights and brawls broke out with little provocation, and shootings were frequent. The town had eight full-time doctors, and it needed every single one of them.

The Hidden Treasure Mine and the Ute-Ulay were very successful operations. . . until the owners decided that all the single miners must live in the company's boarding houses. This was March 1899, and these miners, all Italian union members belonging to the Western Federation of Miners, went on strike. Any worker who tried to return to work was beaten and chased out of town. The mine owners asked Colorado Governor Charles S. Thomas for help. Governor Thomas immediately sent a military unit, along with the Italian Consul, Dr. Cuneo. The military surrounded the town of Henson and waited. The Italian Consul wrote a letter to the strikers. Speaking for the Italian government, the Italian Consul said he would meet with the miners and the government officials, to talk about the problems and peaceably resolve any differences. The miners agreed with the request. When they met for the discussion, however, the miners were arrested immediately. The miners' wives and children were also jailed. Three days later, they were all freed. The area mines, they were informed, would no longer employ Italians. The single men were given three days to leave the county, and married men had sixty days to leave. The mine owners tried to justify their harsh reaction to the strike, explaining that they were trying to control the town, and keep violence to a minimum.

One fateful day during Henson's peak years, every miner's worst nightmare was realized. Twenty miners were working at the Ute-Ulay Mine, and 16 of their counterparts were working at the adjacent Hidden Treasure Mine. On this day, the miners accidentally tunneled into each other, releasing deadly gasses and igniting a deadly explosion. Many of the men were killed, and others were seriously injured. Shock and grief gripped the town.

The town persevered until the mines played out, and the people moved away. By the time the post office finally closed in 1913, no one was left.

SHERMAN
Hinsdale County

Directions: Leaving Lake City, take County Road 30 southwest for about 16 miles. The road will fork into County Road 4 on the right and County Road 18x on the left. Sherman is about one more mile farther between these two roads after the creek.

In the mid 1870's rich silver, gold, copper, and lead strikes were discovered here. In 1877 a mining town called Sherman was formed. It may have been named after General Sherman of Civil War fame, but no one knows for sure. A.D. Freeman was one of Sherman's founders. Sherman was located 12 miles east of Animas Forks and 16 miles southeast of Lake City. It was right at the junction of the Gunnison River and Cottonwood Creek, so close to the water it was actually in the stream bed. North of Sherman, massive Sunshine Peak stands at an elevation of 14,001 feet.

The area had some good mines, but the Black Wonder was the best, with ore worth $50 to $2,000 per ton. The mill was located in town, and operated until 1897. Other mines supporting the town were the Clinton, the George Washington, the Monster, the Mountain View, the New Hope, the Minnie Lee, and the Smile of Fortune. The town had been platted with broad, 60-foot wide streets, and its alleys were 20 feet wide. One huge building downtown on the corner of Main and Sixth streets contained a hotel, grocery, butcher, and bakery. The population peaked at 300.

The Colorado Business Directory for 1878 listed the following on Sherman:

SHERMAN
A settlement on the Gunnison, 19 miles above Lake City.

Grocery and Saloon: H. Deathridge.
Blacksmith: Charles Knapp.
Mining Agent: A. D. Freeman.
Cabinet Maker: J. H. Freeman.
Saw Mill: G. W. Franklin.
Meat Markets: T. L. Lakeman, Spalding & Co.

Within five years, Sherman had grown, with more businesses, more mines, and a population of 125, according to *The Colorado Business Directory* for 1882:

SHERMAN

Mining camp in Hinsdale County, 12 miles east of Animas Forks, and 16 southwest of Lake City. Tri-weekly mails and stages (daily in summer,) to Lake City.

Black Wonder Consol. Min. Co., W. W. Storer, supt.
Coffin, Traver & Co., lumber mill.
Franklin, G. W., groceries.
McIlvain, Harvey, saloon.
Mountain View Mining Co., C. H. Bent, supt.
New Hope Mining Co., G. W. Davis, supt.
Randle, John, gen'l mdse.
Sherman M.& M. Co., G. W. Cory, supt.
Woodruff, C. H., hotel.

Sherman had flooding problems with the spring runoff. One of the area mines started to construct a dam, which they projected to be 150 feet tall upon completion. After the dam was half built, the mine owners filed bankruptcy. Another mining company stepped forward and finished the dam. Almost immediately after construction was completed, a storm rolled in and a cloudburst poured rain until the dam could not hold it any more. The dam burst, destroying much of the town. Sherman was never rebuilt.

Some of the miners returned to Sherman and sporadically worked the mines until 1925.

WHITECROSS
Hinsdale County

Directions: Coming from Lake City take County Road 30 for 16 miles until you reach County Road 304. This is now a rough jeep trail. At this point you are passing the site of Sherman. Drive five miles northwest on County Road 4, which is the route to Cinnamon Pass. This will take you directly to Whitecross, which is just past the town of Burrows Park.

A second route to Whitecross is another difficult jeep road. Take State Highway 110 northeast from Silverton. This will take you past Howardsville, Middleton, and Eureka. You will be driving next to the Animas River. Before reaching Animas Forks, turn right on County Road 5, which goes over Cinnamon Pass. After the Pass the road turns into County Road 4. Stay on this road for a couple of miles. The U.S. Forest Service has put up a sign directing you to the site.

In a high mountain valley meadow, called Burrows Park, a cluster of mining camps formed in 1877, by Cinnamon Pass. The largest of these was Whitecross. A quarter of a mile above Whitecross was the mining town Burrows Park. Tellurium was another camp located just above Whitecross. On the other side of Tellurium was a camp called Sterling. The neighboring camp of Argentum was six miles up Cottonwood Creek.

The town of Whitecross was very hard to reach. It was at the base of Whitecross, which had two large quartz veins in the shape of a cross that stood above the town. In the beginning, its post office was called Burrows Park, but it was changed in 1882 to Whitecross. During the decade from 1890 and 1900, the town had a population of around 300 people.

Close to the timberline, next to Cinnamon Pass, was the Tobasco Mill. It was built in 1901, and it was a hundred-ton mill. The Tobasco Mine was the best producer. The Bonhomme, the Champion, and the Cracker Jack were other important mines supporting the town. The mines in the area produced copper, silver, and galena.

A miner, employed on the night shift in one of the local mines, had stored his explosive blasting powder in a coffee pot in the corner of his cabin. He had a second coffee pot for brewing coffee. Before his shift one night, he put the wrong coffee pot on the stove.

In 1901 the Champion Mill was closed, but in 1916 it was opened and reworked.

The town was very isolated, and the winters were cold. After the silver crash in 1893, the town slowly but surely emptied out.

Today, only a few foundations remain.

ADELAIDE
Lake County

Directions: Adelaide was three miles north of Leadville.

In the year 1876, the town grew up around the Adelaide Mine, a lode discovered earlier that year in Stray Horse Gulch. The town had 28 businesses, including at least four saloons, a post office, a smelter, a mercantile store, engineer, grocer, and 36 cabins. The town even had its own school. In 1882, *The Colorado Business Directory* reported a population of 250.

Some of the key mines in the area were the Adelaide, the Black Prince, the Dolomite, the Eureka, and the Morning Glory.

Senator Gallagher owned two mines in the area, the Mikado and Camp Bird. He was reportedly very hard on his workers. In Oro City, he fell to his death down the shaft of his Moyer Mine. He was said to have been a constant presence, haunting his mines and terrifying miners who worked them. It was difficult for new owners to find workers who were not afraid of Senator Gallagher's ghost.

After about ten years of being a boomtown, the mines started to play out. By 1887, only 44 people were living in Adelaide. The market for silver crashed in 1893, and the town simply faded away.

Today, nothing is left.

BRUMLEY (Bromley Station, Bromley)
Lake County

Directions: On State Highway 82, Brumley is ten miles west of Twin Lakes.

The town began with the name of Bromley Station, Bromley, and then finally, Brumley. It was named originally after a hotel owner in town. The Mount Champion Mine on Mount Champion was Brumley's best producer. Star Mountain and Mountain Boy Gulch were productive mines supporting the community and providing jobs. An aerial tram lifted ore from the mountain peak and brought it down to Half Moon Gulch, where wagons waited to haul it for processing. This three-mile long tram cost $115,000 to build.

Brumley stayed a small town, but it was important because the town served as tollgate for Independence Pass Road, then a toll road. Brumley's population declined when fewer people traveled over the pass.

The only reminders of another era are some concrete foundations and cabin ruins.

EVERETTE (Halfway House, Seiden's House, and Everet)
Lake County

Directions: On State Highway 82, Everette is seven miles west of Twin Lakes.

Established around 1881, there were only two houses here. C.M. Everette came into the area at about this time, and he started to develop the area. This was basically a stagecoach stop on the way to Brumley and Independence Pass Road. *The Colorado Business Directory* for 1882 listed only 15 people living in town for that year.

C.M. Everette promoted the town, and after a few years he had built 30 houses, a few stores, a few hotels and the Everette House, a post office, and a few mills. Everette had at one time some promising mines, two stamp mills, and a water power plant. But soon the mining went dry. Eventually people stopped using the pass, especially when they built a railroad that went to Aspen.

In 1918 W. L. Larimer set up a sluicing operation here and used some of the abandoned buildings.

South of town is where all the mining was done. It was called the Red Mountain District. The area that this district covered was Peek-a-Boo and Sayers Gulches and Lake Creek's south fork. For over 30 years, the mine has been worked intermittently.

Sometime around the 1950's, the last standing building, a stagecoach station, disintegrated.

All that you will find now at Everette are building and cabin foundations.

LEADVILLE
Lake County

Directions: Leadville is on U.S. Highway 24, about 33 miles north of Buena Vista. It has an elevation of 10,190 feet.

Abe Lee was the first man to find gold in California Gulch. It was the winter of 1860, and Lee was panning in the Arkansas River with some other prospectors from Georgia. They had tried their luck at Russell Gulch and then Cashe Creek, but did not find much for their efforts until now. There was snow on the ground, and the sand was frozen beneath the creek. Everything seemed hopeless, and then Abe Lee saw gold on the bottom of his pan.

"I got it! I found it!" shouted Lee. "It's the whole State of California in the 'God damn pan.'" And that is how the gulch came to be named California Gulch.

News traveled fast, and before long, at least 50 more miners were working in the snow. The miners were all in high spirits as they staked out their claims.

Horace and Augusta Tabor were among the newcomers to California Gulch. Horace and the nearby men were quick to erect a cabin for Augusta and her small child. It was very rough. The cabin had no floor, no windows, and no doors. The wagon was used to make furniture, stools, shelves, and a table. They slaughtered the oxen for food. Augusta took in boarders in the tiny cabin, and Horace started to show a profit from his mine before the end of the summer. Horace then established a post office and a general store. Augusta herself weighed out the men's gold with the scales from their store. Horace's mine yielded $5,000 for the summer, but the men below the Tabor mine brought in $80,000 in gold dust. This was a very happy time for these people.

Cabins, tree shelters, and tents were scattered all over California Gulch. The town was first called Boughtown, probably because so many people lived in shelters made of pine tree boughs. When the town grew larger, they called it Oro City.

In just a few months, there were 10,000 men in camp who had come to seek their fortunes. Gambling houses, saloons, and brothels lined California Gulch for miles. One of the prostitutes, nicknamed Red

Stockings, made more than $100,000 within a year. She told everyone she knew that she was going to turn her life around and become a lady. She left town, and no one ever saw her or heard from her again.

The miners would work all day, and they would be paid at night. By the time the sun rose, they had spent all their money, all $6.00 that they had made for the day.

As winter came that year, the gulch emptied. California Gulch had yielded several million dollars in gold in the past few months, but now the miners were deserting the gulch by the hundreds. The miners returned to their hometowns, some went back to their old camps at Clear Creek, and only a few stayed in the neighboring gulches. It was almost impossible to work a frozen claim.

Even the Tabors decided to leave. What use was a post office and store in a ghost town? The claims in the Gulch had played out. The Tabors were surrounded by abandoned cabins and worthless claims. Horace had foolishly staked his claim on the top of a little waterfall, and all the gold from his claim washed down to the claim below. And that is probably why the Tabors' claim yielded only $5,000 in gold that summer, compared with $80,000 for the claim below it. The Tabors packed their meager belongings and left. At this point, with just a few people left in the ghost town, the men tore down the last big gambling hall so they could pan out the gold dust that had lodged beneath the floor. They found $2,000 in gold.

The miners cursed the heavy blackish red sand that kept getting stuck in their sluices. It was actually pure silver mixed with lead, but the miners would not discover what it was for nearly twenty years.

By the winter of 1861, the town was deserted.

The Tabors had heard of a gold strike by two men, Fergeson and Wells, in an area called South Park. Hoping to strike it rich, Horace and Augusta followed the rumor to South Park.

By 1868, the Tabors returned to Oro City and built a small cabin in an isolated area high on the mountainside. Horace had just returned from Buckskin Joe, which was now completely deserted. He was now too discouraged to prospect. He was 40 years old, content to play poker and run his store. They were joined by about 50 other miners. The miners in the gulch were able to pan out about $20,000 in gold a summer, a tiny sum compared with the millions that they had taken out that first summer.

There was one short main street in the town. The town was in ruins, very poor. Horace reestablished his post office and store, and Augusta again took in boarders. There was a saloon or two and a place that served food.

In 1874 William Stevens and A. B. Woods, a miner and a trained metallurgist, came to California Gulch and bought up some mining claims. The two men knew exactly what that troublesome black sand was that kept clogging the miners' sluices. They took samples here and there and traced the source of the silver. The first mine they staked was the Rock Mine, and then the Stone Mine, and then in 1876, the Iron Silver Mine. The Iron Silver Mine later yielded $20,000,000 in silver.

The old miners in the gulch who still hoped to make a strike finally paid attention when Wood sold his half of the claim to Levi Leiter for $40,000. Leiter was from Chicago, and he was the father-in-law to Lord Curzon.

California Gulch stirred with excitement as the miners realized that the mountains surrounding them were filled with silver.

Three poor Irish brothers named the Gallaghers, laborers working a the Rock mine, went over to Stray Horse Gulch, started prospecting, and struck it rich. They named their mines the Camp Bird, the Charleston, and the Pine.

A couple of laborers from the local sawmill sank a shaft at the bottom of Iron Hill, which they called the Carbonate mine. Within three months, they had sold $87,000 worth of ore from their mine.

After hearing about all of the excitement, a much older Abe Lee rushed back to town to find silver in his Dana Mine on Long and Derry Hill.

Tom Wells, who had done very well in the gulch twenty years before, found silver in his Shamrock Mine.

Bill Yankee found silver in his Yankee Doodle Mine and immediately sold it for $50,000.

The Minnie and the A.Y. mines became cornerstones of the fabulous Guggenheim family fortunes. The Guggenheims had mining interests all over the West. Simon Guggenheim, the sixth son of Meyer Guggenheim from Philadelphia, was the newest bonanza king in the area. He soon found himself in the United States Senate.

Other rich mines were the Adelaide, the Catalpa, the Crescent, the Evening Star, and the Morning Star.

The year was 1878, and George Fryer left Fairplay, where he had owned a lunch counter. He was a sandwich vendor. He moved to the Carbonate Camp and bought a claim, which was basically a prospect hole, from "Chicken" Bill Lovel. Fryer had to take in a partner because his supplies were exhausted. He obtained a grubstake for half interest in his claim. At 50 feet down, he struck an iron cap of white-green porphory. This mineral almost always indicates that there is silver beneath, and Fryer found a rich vein of silver. This was now called Fryer Hill, and the name of the mine was the New Discovery.

Four rowdy drinking Irishmen – Pat and Richard Dillon, John Taylor, and Dennis Carter – found a rich strike and named it the Little Chief. This mine was sold within a year for $400.000.

Oro City moved down the gulch so it could be closer to the smelters and new mines. They now called it New Oro City, formerly called Boughtown, Slabtown, and Oro City.

When the city moved down the gulch, so did the Tabors. There were a couple of saloons on the main street already. Horace Tabor set up his general story and post office. He even set up a small saloon attached to his store. Augusta was still taking in boarders, cooking and cleaning and sewing for them. Augusta, said to have a heart of gold, took care of the sick and fed the poor. The Tabors would sell the miners fresh food at cost when they had it.

In January 1878, a group of miners met, proclaimed themselves a town board, and petitioned with the governor to make Leadville a city. With this request, was the request for Tabor to be appointed as Leadville's postmaster. Tabor was also elected mayor for that year. By this time, the town had grown to about 300 miners.

A couple months later, Tabor, who was working in his store at the time, was met by two German miners, George Hook and August Rische, who had given up their shoemaker business in Fairplay to seek their fortunes in

Leadville. They explained to Tabor that they were out of money and supplies and needed a grubstake if Tabor could give them supplies in exchange for being a partner. Tabor did not want any part of it; neither did anyone else in camp. Nevertheless, they kept coming back and after the third time, Tabor just wanted them to stop bothering him, so he agreed to let them take the supplies they needed. On the way out the door one of them grabbed a big jug of whiskey, which had not been intended for their supplies. However, Horace Tabor was a kind man and simply let them go, thinking that he would never see them again, and went back to his daily activities.

Hook and Rische walked about a mile from town. They climbed part way up a small hill and sat in the comfortable shade of a tree. They each took a drink from their jug of whiskey and decided that this was where they would start to dig. After a week of digging and drinking, they had dug about 25 feet down and struck ore. Hook had once worked at the Pittsburgh mills, so he named their strike the Little Pittsburgh Mine.

Rische ran to Tabor's store, waving fists full of ore. "We've struck it!" he screamed. "We've struck it!"

Augusta looked at Rische and told him that when he had money in his hands and not rocks she would believe him.

Tabor and all of Leadville followed Rische back to the spot. Tabor could not believe it. For $64 in supplies, he was rich. He owned a third of this strike called the Little Pittsburgh. Somehow, Hook and Rische hit the top point of a vertical vein, the only area on the whole hill where the vein was anywhere close to the surface, according to a later USGS report.

Now the Tabors sold their store. Augusta was relieved because she had always worked so hard, been so careful. Horace was already 50; now maybe their son Maxcy would have an easier life than they did.

Their new mine was bringing in $20,000 a week. It could have brought in more if they only had more manpower. Hook sold his one-third interest to Rische and Tabor, and then he left for Germany. After Rische had earned $145,000 from the Little Pittsburgh, he sold his stake in the mine for $265,000 to bankers.

In less than a year, Horace Tabor made $500,000 from the mine.

Then, David Moffatt and Jerome Chaffee, Denver bankers, bought Tabor's interest from him for a million dollars cash, plus they gave him stock interests in the Little Pittsburgh, which Tabor sold immediately for another million dollars. Tabor was the first bonanza king in Colorado.

Leadville was attracting miners and people by the thousands. By the summer of 1878, 30,000 people were living in Leadville. There was a shortage of beds for such a great influx of people. The smelters and mills in towns worked nonstop. The smoke of industry lingered in the air. The town was booming.

"Chicken" Bill Lovell was the old prospector who sold George Fryer his prospect hole where Fryer struck it rich with the New Discovery Mine. He had moved to the other side of the hill, which was now called Fryer Hill. "Chicken" Bill had been digging a shaft himself. It was very hard work digging out at the bottom, hoisting out the rubble and dirt, then climbing out of the hole and dumping the bucket. It took a long time, and by the time he had dug his hole 40 feet deep, without any sign of ore or anything of value, he struck water. So, he crawled out of his shaft, by now rapidly filling with water, dried himself, waited for dark, and sneaked over to the Little Pittsburgh's tailings piles. He packed a sack with as much ore as he could carry. He then ran back to his shaft of water on Fryer Hill. "Chicken" Bill threw the ore into the shaft and on the sides. The next day, he visited Horace Tabor in town and explained that he had just made a rich strike 40 feet down, but his shaft flooded out and he did not have the money to pump out the water. He hoped Tabor would be interested in buying the mine. So, Tabor went over to Fryer Hill and took a look at the shaft, the water, and the ore. A few of the men who had accompanied Tabor thought that the ore looked just like the Little Pittsburgh ore, but the did not say anything at the time.

Tabor was impressed, thinking that it looked good. He gave "Chicken" Bill a check for $40,000. "Chicken" Bill took the check to town right away, went to the nearest saloon, and started drinking. Soon drunk and stumbling, he told everyone in the saloon how he had salted the mine with Tabor's own rocks from the Pittsburgh Mine's tailings pile.

The next day, Tabor sent his miners down to the new mine to drain the water. Everyone laughed at Tabor, behind his back at first, but soon they did not even try to hide their laughter.

After Tabor had the men drain the water from the shaft, he told them to start digging. The men dug for three days, and made the shaft eight feet deeper, when they struck the richest lode of ore on Fryer Hill.

Tabor called his mine the Crysolite. He then incorporated the mine, calling it the Crysolite Mining Company. The company was worth $20,000,000.

He then bought a mine called the Matchless for $117,000. He had to pay $40,000 to some other miners for conflicting claims on the Matchless, but it was worth it. It brought in $100,000 a month.

Tabor bought the Vulture Mine for $250,000. He bought into the Maid of Erin Mine for $43,000.

George Fryer sold his half interest of the New Discovery for $50,000 to someone who sold it for $162,000 to Rische and Tabor.

Tabor continued buying up mines and putting money into them, hoping that they would show a profit. Most of these mines were not productive. Some of his mining acquisitions were the Climax, the Denver City, the Dunkin, the Elk, the Empire, the Henrietta, the Hibernia, the Little Willie, the May Queen, the Scooper, the Tam O'Shanter, and the Wheel of Fortune.

Tabor was so busy with all of his business dealings, socializing, and building Leadville that he did not spend much time with Augusta, but she waited home loyally.

On the top of Fryer Hill was an underdeveloped claim called the Robert E. Lee, which Jim Dexter had bought it for $15,000. Dexter hired some miners to help him. They spent a month digging a shaft straight down at least 100 feet. They still saw no sight of ore, after blasting and digging and blasting. When someone offered Dexter $30,000 for his claim, he accepted. He stopped his hired hands, who were at the bottom of his shaft filling a hole that they had drilled out to plant explosive. "Come on up, guys, come up!" shouted Dexter. The miners yelled up that they were ready to "shoot" again, or in other words, they were all set to blow up the bottom of the shaft with gunpowder.

"NO!" yelled Dexter. "Come up. I'm not puttin' any more damn money into this hole."

So, the guys came up, Dexter paid them off, and sent them on their way. And Dexter felt better, knowing that everything was not a loss. At least he was able to recover some of his losses. And then he himself was on his way.

The next day, the owners examined their new mine and decided to ignite the gunpowder that had already been put in place by the previous workers. So they "shot" the prospect hole. After the smoke, dirt, and dust cleared, the new owners climbed down the shaft to take a better look. They could not believe their eyes. There lay the thickest vein of pure silver that they had ever seen.

Within 24 hours, the Robert E. Lee had produced 95 tons of silver ore. That 24 hours cost the owners $60 in labor, but they took out $118,500 worth of ore. The Robert E. Lee was now the top producer of silver in the world.

George Robinson made his fortune much like Tabor. He came to Leadville from Michigan where he had been a banker who had gone bankrupt. Robinson had a little money saved, just enough for him to buy some supplies, build a little shack, and open a store. Some miners came into his store needing supplies, and he grubstaked them. Those dirty, poor miners struck it rich. They named their newfound mine the Wheel of Fortune. George Robinson received millions just for trading supplies for an interest in their claim.

A.V. Hunter and George Trimble made their millions from the Winniemuck Mine.

Horace Tabor enjoyed his money, his status, and the finest material things that money could buy. He bought for himself an exquisite diamond that had belonged to Spain's Queen Isabella. Tabor had decided for business reasons to move his house; so he had his house, which was on Harrison Avenue, picked up and wheeled up the street and dropped right next to the sidewalk on Carbonate Avenue. The house had six small rooms and was simply furnished. Augusta no longer had boarders. There was just Horace, herself, and her son Maxcy, living in the home now. Augusta even had a maid.

Tabor frequently had dinner with his friend Bill Bush at the Saddle Rock Café. One evening, Tabor glanced across the room, and eyes met the soft blue eyes of Elizabeth McCourt, better known as Baby Doe. She had been a dancer in a Central City State Street theater. Some people knew her as Harvey Doe's ex-wife; they had lived together in Central City for almost a year when Baby Doe filed for divorce on the grounds that Harvey did not support her. He could not hold a job; he drank and did not come home. They had moved to Colorado together from Oshkosh, Wisconsin.

Baby Doe had taken a carriage to Leadville from the Central City area. She was very beautiful and less than half Tabor's age. She was sitting alone. She had been watching him. Tabor was an attractive, powerful, self-assured man. He was around 50, with brown hair and a mustache. It was a sight the way people would swarm around him, friends, associates and business owners. Baby Doe smiled at him then turned her head away.

Baby Doe knew that this must be "*the* Horace Tabor." She had seen his picture in the newspaper, read stories and heard people talk about him. Everyone knew who he was. Yes, this was Tabor. Before she had finished eating, Tabor had walked over to her little table, introduced himself, remarking that he had noticed that she was alone and asked her to let him know if there was anything that she needed. This was a very rough place to be alone; it was overcrowded, with miners, gamblers, smelter hands, barkeepers, blacksmiths, and prostitutes. Tabor and Baby Doe started to talk and their friendship began. Tabor asked Baby Doe if he could call on her, and indeed he did.

In November 1778, Tabor had been nominated and elected lieutenant governor of Colorado. To look good as a public figure, he bought a big house in Denver for $40,000 and he spent $20,000 to refurbish and remodel it. When the house was ready, he brought Augusta to it. She seemed distressed. She "could not put her finger on it," but she knew that there was something wrong or soon would be. Tabor now spent much of his time in Denver because of his political responsibilities.

Leadville was growing. By May 1879, there were 82 saloons, 19 hotels, 41 boarding houses, 13 liquor stores, 21 gambling houses, 12 blacksmith shops, 7 smelting works, 6 livery stables, 3 undertakers, 38 restaurants, and 36 brothels.

People said that Leadville never slept. Saloons and gambling houses were open all hours of the might. Most of the men carried guns. Stagecoaches were often robbed at gunpoint. Men were jumping claims, either when the miners were away from their claims or at gunpoint. Desperate men even tried to jump property lots. When a group of men "jumped" the First

Avenue Presbyterian Church, the congregation was forced to abandon the church and had to build elsewhere.

Crime grew worse, and law enforcement was understaffed. For some time there were only four police officers. Then, Officer Bloodsworth was drawn into a drunken brawl with City Marshal O'Connor and shot him to death. Officer Bloodworth then stole someone's horse and left town. The town was down to two law enforcement officers.

Then, Mayor Tabor appointed Mart Duggen as city marshal. Duggan was a tough man, about medium build, known to be a killer and a bully. He liked to draw his gun and show the seven notches. He was as quick with his fists as he was with his gun. He harassed and assaulted regular citizens, but he really enjoyed hunting down criminals, forcing them into deadly gunfights that he was certain to win. Marshal Duggan and Carbonate King Rische were both heavy drinkers. One night they had been drinking, and they started arguing. Marshal Duggan told Rische that he was going to spend the night at "the Pine Street Tombs," the nickname for the Leadville Jail. When Rische refused, Marshal Duggan punched him, knocked him out, dragged him to jail, and forced him to spend the night with a bunch of outlaws in the same small jail cell. Tabor tried to rescue Rische's rescue, but Marshal Duggan looked a Tabor and basically told him to shut up or the same thing would happen to him. Tabor backed down. He realized how much Leadville needed Duggan. Marshal Duggan could not tame Leadville, but it was a start.

Robberies were common in Leadville and its suburbs. Some people were sure that they saw Jessie James and his gang in California Gulch.

Not far from Leadville, a little southwest, is Half Moon Gulch. Stories abound about this place, stories of gold ledges, caves, cliffs, bears, mountain lions, and Jesse James and his gang. Many prospectors had entered the gulch and never returned. Some that made it out later swore that Jessie James and his gang were spying on them from the high ledges of the canyon walls. There was a growing bounty on Jesse James, and a posse of vigilantes quickly and quietly rode into the gulch to capture the men. The vigilantes later said they just missed Jessie James. If Jessie James had any loot hidden there, they said, he would not have had time to take it with him. According to legend, Jesse James' stolen loot is still buried or hidden in Half Moon Gulch.

William H. James was a Welsh watchmaker who had settled in Nevadaville repairing watches until a fire wiped out much of the town. After the fire, he found himself in Leadville working as a partner with Edward Eddy. He later struck it rich and became another of Leadville's bonanza kings. William James was Tabor's successor as mayor of Leadville. Mayor James had to work out a lot of problems. The city treasury was empty. No one would pay their taxes until they were assured that their neighbors were paying their share. The police were unhappy because they had not been paid their $100 a month salary for a long time. There were now eight police officers. People who lived in the lower end of town were upset because at night when they were sleeping, people from the upper part of town were dumping their garbage in their yards. This was the summer of 1879. Despite these squabbles and issues, there were wonderful parties, an archery club, roller skating parties, a bicycle club, and a horseback riding association.

Tabor and his friend William H. ("Bill") Bush, a former math teacher from Kalamazoo, Michigan, spent many evenings together at Tabor's favorite spot, the Clarendon Bar. Tabor was now a pillar of society and owned many mines, a water company, an insurance company, a gas company, a horse car company, two lumber companies, the Leadville Bank, a railway, the Tabor Milling Company, the smelters supply company, and soon the Tabor Opera House. He founded Tabor City above Leadville.

In the winter of 1879, Augusta Tabor left to visit her hometown in Maine. When she returned to Colorado, her relationship with her husband was strained. She was now almost 50 years old. She had to wear thick blue glasses so that she could see, and she did not want to wear fancy dresses all the time. She was more comfortable in the simple dresses that she had always worn.

Horace Tabor abandoned Augusta in January 1881. Before this, he was seldom home with her. He had been, it was rumored, spending his time with 20-year-old Baby Doe.

Augusta Tabor sued her husband. She was still in their house. But when she was finally able to bring the matter to court in Denver, it was January 1883. The trial began quickly. Tabor told the court that he had traveled to Durango, where he had filed for divorce, and for an unknown reason Augusta had not been served papers. In reality, Tabor had paid off a clerk. The entire divorce proceeding was fraudulent. Augusta was forced to accept the settlement of around $250,000 plus the house. Augusta cried in court at the end of the trial when she was given papers to sign. Looking up at the judge, with tears in her eyes, she asked the judge what her name was. The judge reassured her that her last name was Tabor, and she could keep the name. She cried and told the judge to put it on the record that this was not done willingly.

Tabor left for Washington, D.C., three weeks later. He was now a U. S. senator.

On the first of March 1883, Tabor married Baby Doe (Elizabeth McCourt) in an extravagant wedding at the Willard Hotel. The guests at the wedding included the most powerful political men in the country-- President Chester Arthur, Senator and Mrs. Nathaniel P. Hill, Secretary and Mrs. Henry Moore Teller, Senator Jerome Napoleon Chaffee, General Charles Adams – as well as Tabor's friends.

Baby Doe was 22 years old, with blonde hair, blue eyes, and a perfect figure. She was beautiful and charming. The wedding began at nine at night. The bride wore a gown of heavily brocaded white satin. She wore white gloves and carried white roses. Among the many gifts that Tabor gave her, Baby Doe sparkled in a $90,000 diamond necklace. Reverend P. L. Chappelle performed the ceremony. Flowers were everywhere, and violets were strewn on the floor. At the banquet, there was a huge white wedding cake, each dish held a beautiful flower, and champagne buckets were covered with flowers. At the end of the banquet, President Arthur hinted to Baby Doe that he would like a flower from her bouquet. She graciously gave him one, and then at 10:45 the President left. The other guests followed the president's lead and soon left. By midnight the reception room was empty.

Rumors spread about Tabor's fraudulent divorce from Augusta, and that he had married Baby Doe before the showy Washington wedding. People said that he should be brought to court to face bigamy charges.

Maxcy Tabor stood by his mother after the divorce. Augusta lived as a recluse after the divorce. She seldom went out and did not invite anyone to her home. She lost what little money she had in investments in 1893 with the silver crash. She died two years later, on 30 January 1895, the day that would have been her 38th wedding anniversary. She died in Pasadena, and her son Maxcy brought her back to Denver, where she is buried in Riverside Cemetery. Her sister Lilly died in 1925 and was buried next to her.

Colorado newspapers accused Tabor of disgracing the state, disgracing it both in his private life and as a public officer. Tabor became a political and social outcast after he put aside the kind older woman who was his wife.

Tabor showered his beautiful new wife with diamonds, an elegant carriage, and the finest of everything that money could buy. He bought a large house in downtown Denver and had it remodeled into a lavish Italian villa. Horace and Baby Doe were fond of entertaining celebrities and visitors, but found themselves shunned by Denver society.

Soon a baby was born to Baby Doe. They named the baby Elizabeth Bonduel Lillie Tabor. The baby had a wardrobe of 50 robes and velvet gowns and a jewelry box overflowing with rare, expensive jewels.

In 1889, another daughter was born, and they named her Rose Mary Echo Silver Dollar Tabor. She, like her sister, wore an $800 christening robe.

Tabor started to have problems. His mine started to become worthless. It had been a long time since he had made a profit, and now he had spent hundreds of thousands of dollars to see if he could save his holdings. His efforts were unsuccessful. Tabor had just lost $500,000 on the Calumet Dock project on Lake Michigan. He speculated in the commodities markets and lost $500,000 in the grain pits.

Tabor's lumber, gas, water, and realty companies in Leadville became worthless, and then started to cost him money as the camp fell apart. His Texas copper mines proved to be of no value. His fire and life insurance companies were no good, so he dissolved them. Millions of his dollars had been wasted in politics. And, it was said he started losing huge sums at poker.

In 1893, the great panic struck. In Denver, 12 big banks closed in three days. The miners lost their biggest customer for silver, the U. S. government.

One of Tabor's last efforts to regain his fortune was to work a mine called the Eclipse up Boulder Creek.

Tabor went to go see the carpenter he had hired years ago in Leadville, Winfred Scott Stratton. Stratton had now made millions on mining prospects in Cripple Creek. Tabor explained his situation to Stratton, how he needed a little capital to work the Eclipse Mine, and maybe open the Matchless again.

Stratton wrote a check to Tabor for $15,000. Tabor was surprised and relieved. Of all the people that Tabor had helped, only Stratton was there for him during these desperate times.

Tabor went to work at the Eclipse. His money soon ran out, and he had not found anything. He lost the Opera House, his Italian villa, his carriages and stables, and all the family diamonds. Tabor was almost 70 years old, penniless and desperate.

It was January 1898. Tabor had a very hard year. Senators Wolcott and Teller were able to secure an appointment in Denver for him as a postmaster. He now had a job that paid a salary of $3,500 a year, and he had enjoyed being a postmaster many years before in California Gulch. Tabor, Baby Doe, and their two daughters shared one small room in the Windsor Hotel.

On 10 April 1899 after Tabor had been feeling ill for a week, he died in his hotel room. Baby Doe, Silver Dollar, and Lillie ("Golden Eagle") were at his side. Maxcy Tabor had been sent for, but he did not make it in time.

Horace Tabor was buried at Jack O'Neill's Ranch, now called Mount Prospect Cemetery. "Tabor" is the only thing written on his small granite stone.

Two months after Tabor's death, his family was living in poverty. For a while Baby Doe and the young girls lived in a little brick house off Larimer Street.

Baby Doe returned to Leadville, taking her daughters with her. They moved into an old cabin next to the Matchless Mine shaft. The cabin was a little over a mile above Leadville, up Stray Horse Gulch on Fryer Hill.

After years of living in poverty on the Matchless, the older daughter Lillie went to live with her grandmother in Chicago. She had been arguing with her mother, who was now in her 40's and destitute. Lillie soon met a young man at Grandmother McCourt's. She married him, moved to Wisconsin, and never spoke to her mother again.

Baby Doe was upset, but she still had Silver Dollar. Every day she inspected the Matchless, checking ore samples and old machinery. She often climbed down into the mine, descending the 300 feet to see how high the water was that had flooded the shaft and tunnels.

Silver Dollar had tried to support her mother and herself by writing poems and short stories. Her efforts were unsuccessful, so she moved to Denver and tried to write for *The Denver Times*. That did not work, either. A few years later, Silver Dollar, who was now just over 30 years old, told her mother that she was going to enter a convent, and she left. Baby Doe never heard from her again. Baby Doe was now alone. Her husband had died years earlier, and both daughters deserted her.

Baby Doe had been visiting in Denver, walking that evening in the street she heard the news. Baby Doe did not have the few pennies that she needed to buy a newspaper. She walked three miles to the library. By now, it was late, and The Denver Public Library was closing. One of the employees took a close look at her. She was dressed in rags and wearing mining boots that did not fit, and was very intent on seeing the paper. He let her in. Silver Dollar's picture was on the front page of the paper, and the story was under her picture. Silver Dollar had died in Chicago under mysterious circumstances. It was said that she fell into a tub of boiling water and was badly burned by scalding water. She died a few hours after this happened. She insisted that it was an accident.

Baby Doe choked back her tears, and cried, "Silver." She stumbled out of the library.

No one claimed Silver Dollar's body. After a while, Baby Doe's brother Peter McCourt mailed money to have her buried. Baby Doe did not go to the funeral. She denied that her daughter Silver Dollar was the wasted woman who had died such a horrible death in Chicago. This older woman had been an unknown prostitute and drug user. Her daughter Silver Dollar, she insisted, was healthy and happy in a convent.

Sadly, Baby Doe returned to her cabin in Leadville. She still had the Matchless. She had lost everything else. One of the last things that her husband told her to do before he died was to "hold on to the Matchless."

Living in her little cabin, she had few friends and kept to herself. Sue Bonnie was her closest friend and her closest neighbor.

It was early March 1936. The temperature had hovered around zero after at least a week of severe snowstorms. It had been almost two weeks since anyone had seen Baby Doe. Covered in mud and ice, she had stumbled into Leadville. She had to walk through the high snowdrifts from her cabin down to the road that goes through Stray Horse Gulch into town. Needing supplies, she had been seen at the local Leadville grocery store, buying stale bread and some meat to boil. The delivery boy for the store was the last person who had seen her. He had given her a ride up Stray Horse Gulch to her cabin. Baby Doe had been using gunnysacks wrapped around her feet for shoes.

Sue Bonnie looked up Fryer Hill, and she noticed there was no smoke coming from the shack. It had been a very cold March. Sue Bonnie was worried. She found their mutual friend, an old prospector named Tom French, and the two hiked up the snowy path to Baby Doe's cabin. There was no smoke coming out of the chimney. There was no movement or sound. Looking through the window they saw a motionless Baby Doe lying on the floor. Her arms were outstretched, and her eyes were open looking up towards the sky. Baby Doe had slowly frozen to death. The stove had been out for a long time. A pile of wood was stacked in the corner of the cabin. The cabin was a mess. Her battered old iron bed was in the far corner. Newspapers, bundled up letters, and boxes were strewn across half of the room. Boxes had been used as shelves.

Baby Doe was buried in Denver's Mount Olivet Cemetery.

MALTA (Swill Town)
Lake County

Directions: Malta is four miles southwest of Leadville on U.S. Highway 24.

The little town of Malta was founded in 1871 at the front of California Gulch. Malta boasted the first smelter for the area built in 1875.

By 1878, Malta had grown into a community of 300 people, with at least 50 houses, a post office, 4 hotels, and a brewery. The following year, in 1879, a racetrack was built at a cost of $5,000. It was used by the Carbonate Kings of Leadville. The peak of Malta's life was in 1881, when there were over 300 residents.

Malta still has some residents, and a railroad junction passes through the town.

Two miles from Leadville on U.S. Highway 24 is the Leadville National Fish Hatchery. Established in 1889, this hatchery is one of the largest in the country. Free hatchery tours are available, and nature trails are nearby.

ORO CITY and NEW ORO CITY (Agassiz, Kelly's Diggings, Poverty Flats, Boughtown, Shaptown, and California Gulch)

Lake County

Directions: Go to Oro City by taking the road up California Gulch for three miles from Leadville.

It was in Oro City that Colorado's Gold Rush was ignited in 1860. This was an important booming town that drew thousands of miners. So many small camps were built that these camps finally grew into one big camp until Leadville was born. Nothing is left of these early camps, just a few paragraphs in history.

During World War II, the men in the area used the old machinery and mining equipment for their scrap metal drive. People in the area used the wood from the old buildings for firewood.

The Colorado Business Directory for 1882 listed the following for Oro City;

ORO CITY
Mining camp in Lake County, 3 miles east of Leadville.

Ringold Bros., stamp mill.
W. H. Stevens Mining Co.

By 1896, the town had grown in population and boasted saloons and a general store. *The Colorado Business Directory* listed the following on Oro City for that year:

ORO CITY
Old town in Lake county, 3 miles from Leadville, the nearest railroad station. Population 300. Altitude 11,000. No post office. Mail to Leadville.

Caulfield, W J postmstr
Collier, Phil, saloon
Hicks, Thos, saloon
Kelly, John, saloon
King, John, gen mdse

Wood, S, pres, Lillian & Antioch Mining Cos

The Carbonate Weekly Chronicle reported the conflict between two French miners in 1880:

WAR IN ORO.
Two Frenchmen Attempt to Eat Each Other Up.
Among the usually quiet and peace-loving denizens of the camp of Oro are two French miners, who have long cherished an ancient grudge against one another. Their threats have been loud, deep and awful, and it was generally understood about the diggings that when they met some bloodletting would take place. Last evening this interesting event occurred. Frenchmen No. 1 was passing a butcher shop carrying two pails of water when No. 2 suddenly sprang out and dealt the astonished Gaul a blow upon the nose, following it up with a general assault at arms. The assaulted miner fled. Waterloo was nothing to it, and tearing to his house returned brandishing an enormous navy revolver and filling the air with French oaths four stories high. The pugilist might have withstood the infantry, but when the military was unlimbered he lost no time in getting out of range. At this juncture Dave Bingham, the peace officer of the camp, put in an appearance, and seizing the weapon demanding a cessation of hostilities. Meantime the town had rushed to the seat of war. Miners had left their claims, teamsters had run from their wagons, mothers left their babies falling down stairs and everybody hurried to see the edifying spectacle of one Frenchman killing another, and were hugely disappointed that the massacre was not allowed to proceed. Thus the war stands, but both swear vengeance and the end is not yet.

The Carbonate Weekly Chronicle
28 August 1880

SAINT KEVINS (Sowbelly Gulch, Amity)
Lake County

Directions: Saint Kevins is seven miles northwest of Leadville.

In the mid 1880's, Thomas Walsh, an employee of the Colorado Central Railroad, decided to try his luck at prospecting. He and his wife lived in Sowbelly Gulch when he made a silver strike, calling it the Saint Kevins Mine. After that, the Walshes changed the name of Sowbelly Gulch to Saint Kevins. Walsh also helped to plan out the town. They built a school, mill, a boardinghouse, and store. Just south of town is Turquoise Lake.

The Saint Kevins Mine is where the camp was built. The other mines were the Amity, the Griffin, and the President. Because some of the miners who worked in the Amity Mine lived around it, it could be classified as a mining camp.

Following the Great Silver Panic in 1893, nearly everyone left.

STRINGTOWN
Lake County

Directions: Stringtown is one mile southwest of Leadville on U.S. Highway 24.

Stringtown was founded in 1879 as one of the suburbs south of Leadville in the middle of Jacktown and Bucktown. Between Stringtown and Bucktown was the Arkansas Valley Plant, which employed many of the people in the community. There were a few other smelter plants in the area also. There are a couple of industrial sites in use today. Stringtown had a few houses, cabins, and boarding houses – even a hotel called The Great Northern. There were a couple businesses and a few saloons.

STUMPTOWN
Lake County

Directions: Go to Stumptown by taking East Seventh Street from Leadville and going up South Evans Gulch for about three miles. It is located in a shallow valley.

Stumptown, another suburb of Leadville, was named for Joseph Stumph, one of the early settlers. In 1879, Stumptown began as a camp when silver was discovered. The Little Ellen Mine was the first mine to produce carbonate ore. Soon to follow were the Bould, the Little Bob, the Louise, the St. Louis, and the Winnie. There were about five saloons, a few dozen homes, and a large pool hall that featured gambling.

The camp was booming until 1893 when the silver crash hit. In 1895, the camp temporarily rebounded, and finally was deserted in the 1930s.

TABOR CITY (Tabor, Taylor City, Halfway House, and Chalk Creek Ranch)
Lake County

Directions: Tabor City is on State Highway 91, ten miles north of Leadville.

In 1879, Colonel Taylor purchased some property on Chalk Creek, today the site of Chalk Creek Ranch. Colonel Taylor laid out the plans and started his town, Taylor City. The site was developed because Chalk Creek had been a good mining site. Between 1879 to 1880, the community exploded from one cabin to numerous cabins, a general store, a couple of hotels, a blacksmith shop, a couple of restaurants, a livery stable, and a post office. The population reached 150.

Horace Tabor, although living in Leadville, had a stake in one of the mines here, the Little Pittsburgh Mine. The miners idolized him. Tabor was now living the American dream. The miners in town decided to rename their town Tabor City. So Tabor City it was.

Halfway House was either part of Tabor City, or it could have been another small camp adjacent to Tabor City.

Tabor City was indeed a mining town and Halfway House was used as a supply and rest station by travelers going over Freemont Pass. Unfortunately, the ore found in the area was very low grade. People started moving away. By 1881 the post office closed. By the mid 1880's the area turned into a ghost town.

Today, the area is ranchland.

TWIN LAKES AND DAYTON
Lake County

Directions: From Leadville, take U.S. Highway 24 south and turn on Highway 82 right to Twin Lakes. Twin Lakes is 21 miles southwest from Leadville.

Twin Lakes still has a small year-around population and a post office; it is also a summer resort. Before there was Twin Lakes, there was a mining camp named Dayton that was established in 1863. Dayton was the biggest town in the county. For a little while it was even the county seat. The town sprawled from the base of Mount Elbert to the present site of Twin Lakes.

In 1863, the Ryan House was built. It was a stagecoach stop. Besides being a busy mining area, many stages drove through this area going from Leadville to Aspen. Stagecoaches would have to go over Independence Pass, then called Hunter's Pass. Ryan's House is now Colorado's oldest stopping place. It has been renovated, and it is now a beautiful house with white shutters.

In the late 1860's, the two lakes in the area were turned into a small resort. By 1867, the population had grown to 500. This beautiful area drew the wealthy elite from Leadville. They started to call the area Twin Lakes in 1879, and by 1883 the area was surrounded by cabins and six or seven hotels. The largest hotel was called the Interlocken. Rooms were $4 a night, very expensive for the time. The hotel boasted that its cuisine was the best in the country. It featured an ice skating rink in the winter and a big dancing pavilion in the summer.

In the early 1880's, Twin Lakes grew, and it grew around Dayton. Actually, Twin Lakes engulfed the town of Dayton, and soon the entire area was called Twin lakes.

Carbonate King John Campian built a $100,000 house here for his bride. He traveled to Europe on his honeymoon and bought furnishings for his home. He convinced many other carbonate kings from Leadville to build summerhouses here.

Twin Lakes was everyone's resort until a company calling itself the Twin Lakes Resort Company moved in and built a dam between the lakes. The company was able to control the level of the lake water, raising it and lowering it at will. This outraged many of the residents, who moved away in frustration. The lakes were never the same. John Campian moved away and sold his home to someone who turned it into a hotel, the Campian Hotel. Many beautiful old buildings are here, well maintained and inhabited.

Early miners kept the smelters away from town. The mining area stretched from Mount Elbert to Cashe Creek, which was south and then over Independence Pass. The miners also mined the Red Mountain District and Mount Champion. The Gordon Claim was one of the most productive mines. Established in 1880, it even had its own mill. Over the period of 35 years, this mine produced $500,000. Most of the ore in the area was low grade.

La PLATA CITY
La Plata County

Directions: Leaving Durango from the west side of town take U.S. Highway 160 west for about 11 miles. Then take County Road 124, which will be on your right going north for about 7 ½ miles. La Plata will be on your left, just off the road, between Forest Route 344 and Forest Route 061.

A small town formed in a canyon below, and on the eastern side of the La Plata Mountains. It was in 1875, and it was on the western banks of the La Plata River. A century before, Juan Marie de Rivera had led a large party of Spanish explorers through the area searching for gold. In 1875 rich gold and silver strikes were discovered in the area. The town that formed about 2 ½ miles to the north of Parrott City was called La Plata, which means silver in Spanish. Good mines were soon located in the area. The first mine was the Comstock, the second was the La Plata, then the Lady Elenor, the Cumberland, and then the Snowstorm, the Gold King, the Neglected, the Red Cloud, the Swamp Angel, the Mayday, and the Idaho.

By 1882, a post office was established in town, and the town soon had 200 residents. The population peaked at 500.

In the beginning, it was mostly silver that was mined. With the silver crash of 1893, the town came to a standstill, and prospectors switched their efforts toward gold. Eight miles south of La Plata on the La Plata River was a little coal mining town called Hesperus. A stagecoach ran between the two towns daily. La Plata had a main street of false-fronted buildings, stores, boarding houses, and a stage station. La Plata never had a railroad.

By the turn of the century the area mines stopped producing except for the Egyptian Queen, which was renamed and worked for another 50 years. It just was not enough to support the town, and it was abandoned.

Today, a few people live in the area.

PARROTT CITY
La Plata County

Directions: Leaving Durango from the west side of town, take U.S. Highway 160 west for about 11 miles, then take County Road 124 for 5 miles. The site of Parrott City will be on your left, in a field.

In the mid 1850's, Captain John Moss led a party of California prospectors through the area. They explored the mouth of the La Plata River, finding gold in the sands, and then returned to California. When Captain John Moss and his men returned in 1873, the area was still Ute Indian Territory. To gain access to the area, Captain Moss negotiated a treaty with Ute Chief Ignacia, trading 100 ponies and a wagonload of blankets for the right to mine and farm in a 36-mile radius. Captain Moss then went back to San Fransisco with tellurium and gold samples to show a rich banker named Tiburcio Parrott. The ores seemed very promising, and Tiburcio Parrott gave Captain Moss the financial backing that he needed. By May 1874 Captain Moss and his party made their way back to the mouth of the La Plata River. They were surprised to see a group of prospectors from Arizona working their spot. Fortunately they were able to split the 36-mile area in half. Soon, they were all working together, digging irrigation ditches to bring water to their sluices. Captain Moss and his men then platted Parrott City, naming their town after Tiburcio Parrott. By 1876 a sawmill was erected, and well-built buildings and homes filled the area. Good mines were established in about 500 locations, although only 20 mines were developed. The mines in the area produced gold and silver. The best ones were the Ashland, the Bulldozer, the Comstock, the Isabel, the Snowstorm, and the Tenbrook. The ores assayed at $50 to $2,000 a ton.

In 1882 *The Colorado Business Directory* listed the following information on Parrott City.

PARROTT CITY
Mining camp in La Plata County, 16 miles northwest of Durango, and 20 south of Rico. Population, 100. Daily Mails.

Clark, Mort & Co., gen'l mdse.
Finley, W. F., saloon.
Hawthorn & Root, livery.
Hickman, Henry, attorney.
Miners Hotel.
Ordway, W. F., hotel and saloon.

Smith, H. M., builder.
Walters & Lee, meat market.
Winters, W. K., drugs, physician and postmaster.

By 1885, *Crofutt's Grip-Sack Guide of Colorado* reported that the population had grown to 250. It also reported that agriculture and stock raising were the main occupations west and the south of Parrott City.

By the late 1880's, the mines stopped producing, and everyone left. The site of Parrott City turned into ranchland.

Only foundations remain, hidden in the underbrush.

BACHELOR (Teller)
Mineral County

Directions: From Creede, take County Road 504, for four miles. The road to Bachelor will ramble west and then north.

Some prospecting has been done in the Bachelor Mountain area since 1885. The Bachelor Mine was on Bachelor Mountain next to the town of Sunnyside. Sources disagree on who discovered Bachelor Mine; John C. Mackenzie, J.B. Burrett. C.F. Nelson, George Wilson, and James Wilson are all credited with the find. The Bachelor Mine was a good mine, but it was not productive enough to justify building a town.

In August 1891, a prospector named Theodore Renninger was one of a group of prospectors on Bachelor Mountain. They noticed a rich outcropping. The men had the ore assayed, and learned that it was worth $170 a ton, and that was the value of ore on top of the ground. They had yet to start digging. The same men had just found the New York / Last Chance Mine. The men sold the mine immediately for $20,000 to the Bachelor Mining Company.

In September 1891, homesteaders C.L. Calvin and his wife built a home and a boardinghouse where Bachelor would be platted out the following year.

Nicholas Creede had been prospecting in the area when other men discovered the New York / Last Chance Mine. He had discovered some rich float himself, but he waited until the men at the New York / Last Chance had put up their claim sign and their stakes. Then Nicholas Creede filed a claim on the adjacent property north of the New York / Last Chance. He called it the Amethyst Mine. Its ore assayed at $170 a ton, the same value as ore from the neighboring New York / Last Chance.

As news of these discoveries spread, hundreds of hopeful prospectors flooded the area. Curiously, the ore from the Amethyst and the New York / Last Chance was a type that did not need to be sorted, so there were no tailings piles.

An eighty-acre town site was platted. Two saloons and a parlor house went up immediately. Property had been set aside so that a schoolhouse could be built. A lot of people were in the town already building houses. There were many children. Squatters started to stake claims on the school property, saying that there was minerals underneath the ground, and that the government needed to let them mine it. This infuriated many of the townspeople, especially the families who objected to squatters claiming the school property. The town was divided, with nearly everyone vehemently taking one side or the other. News of a potential town war reached Denver. In January 1892, the Colorado State Land Board and the Governor visited Bachelor to speak to the community. The squatters relinquished their questionable claims to the school property, and peace was restored.

In January 1892, Bachelor had its first fire. The Lundy Saloon, Sherry's Saloon, and the Free Coinage Hotel went up in flames. Another fire later in 1892 destroyed another saloon, and then they had a third fire. The town had an excellent fire department and bought their own fire engine at a cost of $675.

In April 1892 when the post office for Bachelor was established, the town had to find another name because there was already a town named Bachelor in California. The town picked the name Teller for its post office, but residents still called their town Bachelor.

The population peaked at 6,000. Bachelor had community dances and musical benefits were held in the Bachelor Opera house. The Bachelor City Drama Club held plays. With a smiling Father Downey leading their efforts, the town built their own Catholic Church.

Early in the year 1893, Bachelor was a boomtown, with everything going for it. Miners were working round-the-clock shifts in the town's 20 mines. On Bachelor Mountain alone there was the Amethyst, the Bachelor, the Cleopatra, the Commodore, the Del Monte, the New York / Last Chance, the Spar, and the Sunnyside.

Later that year, the silver crash hit, and the town never recovered.

The Colorado State Directory for 1896 listed the following on Bachelor:

BACHELOR (Teller P. O.)
The center of the miner population of Creede camp, two and one half miles northwest of Creede. Population 800. Altitude 10,100.

Bachelor Grocery Co., F. Gillett, manager.
Barnett, H M, architect and builder.
Barnie & Rice, groceries and meat market.

Bay & Co., Geo. W S, Last Chance boarding house.
Belcher, A C, blacksmith.
Biles, J. A, physician.
Broade, E E, saw mill and lumber.
Bruner, Mrs. Rosa L, millinery and dressmaker.
Charlton, Mrs. W, boarding
Colville, John B, gents furnishings.
Crawford, E H, supt. New York Chance Mine.
Cunningham & Sloan, saw mill and lumber.
Doherty, J H, saloon.
Eades, A B, dry goods.
Heilscher, August, jeweler.
Henneberg, Misses, dressm`ks
Hughes, W M., blacksmith.
Irwin, Rev. John, pastor Congregational Church.
Jenkins, J W, saloon.
Lee, C B, barber.
Likins, F L, drugs and jew`ly.
Loucks, W. transfer.
Lynch, Geo., confectionery.
McDonald, A K, blacksmith.
McDonald, Dan, marshal.
McKenzie, D B, mch`t. tailor.
McLeod, R J, blacksmith New York Chance Mine.
McPhee, Angus, saloon.
MacLean, A, merchant tailor.
Mallon, H P, saloon.
May, I S, gen. mdse. and meat market.
Manion, Mrs. M, boarding.
Markley, Tailor, blacksmith New York Chance Mine.
O K Clothing House, A C Monday, manager
Page, J B, assayer New York Chance Mine.
Pollock, Mrs. Laura J, teacher public school.
Ray, Ed M, supt Unibell Leasing Co.
Reilly, A F, town clerk and constable.
Ridenour & Henry, livery.
Rodman, Julius, resident agt. New York Chance mine.
Swords, Mrs M, Bachelor City Hotel.
Terrell, J K, clothing, boots, shoes and furnishings.

Van Norden, S E, postmaster, att`y and notary.
Van Norden, Mrs S E, news, confect`y & cigars.
Vincent, T W, coal & lumber
Vincent`s Opera House, Levi & Lewin, props.
Walsh, Mrs John, boarding.
Warren & Coulson, furniture.
Weinrich, W, cabinet maker.
Wells, E D, gen`l mdse.
Whitehead, J H, just. peace.

The townspeople stayed on until the turn of the century, trying to keep their town and way of life alive. Finally, the residents had to leave so that they themselves could survive.

Today, only ruins remain.

SUNNYSIDE
Mineral County

Directions: To get to Sunnyside, take Highway 149 north of Creede for three miles. Then take an unmarked road west. This road will double back around the mountain edge and then up the gulch, where you will see some tailings piles. Sunnyside is on the other side of the mountain west of Bachelor. An old road linked Bachelor to Sunnyside.

Sunnyside was an older camp formed in the early 1870's following some small discoveries. In the mid 1880's, the Bachelor Mine and the Alpha Mine were discovered. In 1885, Dick Irwin came to Sunnyside with financial backing from some West Coast investors. He discovered and developed the Diamond Lode, the Hidden Treasure, and the Nelson. The Sunnyside Mine was a profitable mine.

In 1887 a post office was opened in Sunnyside. In the early 1890's Sunnyside was doing well and more mines opened. The Corsair, the Kreutzer, the Sonata, and the Yellow Jacket mines were producing well.

Transportation was always a serious problem for Sunnyside.

Today, nothing is left of Sunnyside, and the site itself is very hard to locate.

GUSTON
Ouray County

Directions: Coming from Ouray, go south on U.S. Highway 550 for ten miles, than take an unmarked dirt road on the left, a mile and a half before Red Mountain. Take this dirt road for about a mile until the road ends.

In the summer of 1881, John Robinson and a small group of prospectors discovered the Guston Mine. The ore was low grade, and they thought the mine was worthless, but a worker from the Pueblo Smelting Company urged them not to give up. They continued working in the mine and discovered a lode of rich silver ore. The Guston Mine produced hundreds of thousands of dollars in ore.

The following year, in 1882, John Robinson made another discovery. He had been walking around the Guston Mine, and was only a few hundred feet from it, when he found a sample of almost pure galena. Robinson staked his claim on it, naming it the Yankee Girl. He put down a twenty-foot shaft and then staked claims on either side of the mine, calling them the Robinson and the Orphan Boy. John Robinson sold the Yankee Girl for $125,000 to finance the development of his two latest mines. Both the Robinson and the Orphan Boy were very successful.

Guston, named for the Guston Mine, grew up between Red Mountain and Ironton. The mines in the area were worked by the miners from all three towns. Guston's population had averaged around 300, but soon grew to about 1,000 in the late 1890's. Many more mines were found in the area such as the Canice, the Genessee, the Paymaster, the Saratoga, the Silver Bell, and the Vandervelt.

In 1891, the Denver Congregational Church sent Reverend William Davis to the Red Mountain District to build a church and to start a congregation of his own. When the minister went to the town of Red Mountain, no one wanted anything to do with him. So he and his family moved on to Guston, where the town welcomed him. He told the citizens how much he wanted a church, and people in the district donated money and labor to make his dream come true: Lt. Governor Story donated the land; Ouray's Reverend Gaston donated the pews; residents of Guston donated $300; the men in town donated $30 and also provided much of the labor to build the church; the town of Silverton gave $100; and the town's Cornish miners made and donated a cupola, a bell, and a whistle. This was the only church in the Red Mountain District. It was built on the crest of a hill just outside of town, and it stood there, facing the vicissitudes of weather and ore prices until it fell in the mid 1940's.

After the turn of the century the townspeople moved away to more prosperous towns. Mine production slowed. With newer equipment, fewer men were needed to work the mines and mills.

Today, nothing is left of Guston.

IRONTON   (Copper Glen)
Ouray County

Directions: Leaving Ouray, go south on U.S. Highway 550 for about 6 ½ miles. The town site of Ironton is on a dirt road that runs parallel with the highway on the left side of the road. A few dirt roads turn left off the highway into the trees and onto the main street of Ironton. It has an elevation of 11,018 feet.

In 1882, after the discovery of rich lead and silver ores, the little town of Ironton formed about three miles north of Red Mountain Pass. In the beginning it was also called Copper Glen, but eventually everyone called the town Ironton. The town began as a tent city with 125 residents, according to *Crofutt's Grip-Sack Guide of Colorado*. The men worked at the nearby Guston and Yankee Girl mines. A few months after the town was settled, more permanent buildings went up. Ironton's neighbors to the south were Guston, one mile southwest, and Red Mountain Town, two miles southwest. Guston and Ironton were about the same size, but Red Mountain Town had a population of 1,000. Red Mountain Town was the "happening place to be," with gambling, saloons, and entertainment 24 hours a day.

In the late 1880's the Otto Mears Rainbow Route Railroad was extended from Silverton, which was 13 miles south, to Ironton, making Ironton the supply center for the Red Mountain District. Ironton grew and had been doing well until 1893 when the silver market crashed. After 1893, many of Ironton's residents left.

A brief advertisement for The Ironton Hotel appeared in Ouray's *Solid Muldon* newspaper on 23 September 1887:

THE IRONTON HOTEL
W. R. Morrison, Prop.
Ironton, Colorado.

The Tables are always Supplied with
the Best the Market Affords.

In the late 1890's gold was discovered in the area, reviving the town briefly.

The Colorado State Business Directory for 1896 listed the following on Ironton:

IRONTON

Station on the Silverton Ry, in Ouray county. Mining the chief industry. Population 250.

Blumberg, A D, meat market
Cambell & Carr, groceries & hardware
Carr, John E, postmaster
Commons, Prtrick [sic], saloon & restaurant
Jefferson, Wm, saloon
Lyon, O P, lumber & sawmill
Roos Bros & Co, groceries & miners supplies

By the 1930's Ironton was empty.

As of the fall of 2000, no business buildings were left in Ironton. The only survivors are six houses hidden in the trees off the main street of Ironton.

This is one of many abandoned houses in Ironton. Today, the only gold in this abandoned community is the spectacular gold of its aspens in autumn.

RED MOUNTAIN TOWN
Ouray County

Directions: Coming from Ouray, take U.S. 550, nicknamed "the Million Dollar Highway," south for eight miles. This will take you over Red Mountain Pass at an elevation of 11,018 feet. After going over the pass, on the northeast side, and after driving a mile, a dirt road leaves the regular road going east for almost a quarter of a mile. This leads to a meadow in which you will find the ruins and foundations of Red Mountain Town.

In 1879, rich silver strikes in the Red Mountain District caught the imaginations of many of Colorado's prospectors. Because the area was so remote and the winters were so harsh, men did not come to the area until 1882.

In February 1882 *The Summit County Circular* reprinted an article on Red Mountain that had originally appeared in *The Leadville Chronicle*:

Red Mountain

The Leadville Chronicle of a recent date says: Among the mining districts that promise to come into special prominence during the approaching season is the Red Mountain. It is situated on the stage road between this city and Independence, about two miles and a half beyond Everette postoffice, and some eight miles from Twin Lakes. The first discoveries were made in the district during the winter of 1879: and its population consisted of just four men. During the year of 1880 there was a good deal of prospecting, and many claims located, but only work enough was done on them to hold them. Only during the past year has there been any development work done. There are now altogether about one hundred locations, some thirty or forty show pay mineral. There is now a ten stamp mill at Everett, and it has plenty of ore to start on, but for some reason remains idle.

In 1882 prospectors discovered a large cavern, which was completely lined with glittering lead carbonate. The founder named it the National Belle. The men began shipping between 50 and 90 tons of ore a day. When the ore was exhausted, the men dug deeper and discovered a large deposit of copper.

Soon, a second large deposit of copper was found and named the Enterprise.

At the same time, the Summit and Congress mines were discovered. These rich strikes were found in the middle of winter.

Cold weather and winter blizzards did not deter prospectors who had heard about this strike and earlier finds. By January 1883, a tent city appeared from nowhere in the snow-filled meadow. By March 1883, an entire town had been built, with hotels, saloons, restaurants, dancehalls, general stores, meat markets, and even a telephone office. The town was named Red Mountain Town after the massive Red Mountain, known for the red stains streaking its sides.

The area mines spurred the financial interests of eastern capitalists who were willing to invest large amounts of money in the area mines. In 1883, Otto Mears built a toll road into the area. The population grew into the thousands.

Before the winter of 1886, the entire town of Red Mountain packed up and moved from Congress Hill, to a site half of a mile away. Some say that it was to be near the new toll road, but others say it was to be near the liquor store and saloon that had recently opened for business. The town's new location was a flat near Red Mountain Creek. Over 50 buildings were built almost immediately, and a waste works system was built for running water.

In 1887, a weekly newspaper, *The Red Mountain Journal,* started publication.

There were now reportedly 10,000 people living in Red Mountain, and at least a hundred businesses. Red Mountain Town was a rowdy, rough town. Saloons were open around the clock. Red Mountain had its own entertainment, dances, clubs, plays, and musical performances.

By the fall of 1888, Otto Mears' railroad was open and running from Silverton to Red Mountain. It was called the Silverton Railroad Company's Rainbow Route, boasting unequaled mountain scenery for its "trip around the circle" between Silverton and all points in the famous Red Mountain Country. The railroad solved major transportation problems, but the tracks had to be constantly cleared of deep snows in the winter. Every time there was a snowfall, 50 men at a time were sent to shovel snow off the tracks.

The Red Mountain Journal printed the following article on Friday, 27 March 1891 about the Yankee Girl Mine.

The Silver District.
The Yankee Girl Mine of To-Day and the Future.
Richest Mine in the world.
Tons of Rich Stromeyerite in the Stopes and Drifts- A Greater Producer Than it ever Was Before.
From the *Ouray Plaindealer.*

The richest ore in America, if not the richest mine, may be truly said of the Yankee Girl, the great Red Mountain producer. In the history of mining there is not found a mine which parallels the Yankee Girl in extent of rich ore. The property has been operated eight years, and could have been a producer every day during that time, but there was a period in history of the mine when no ore was produced, owing to a system of exploration which departed from what has been established as the natural trend of ore bodies. On returning to the course followed by the ore from the discovery shaft, not only was production restored, but the output was largely increased in both extent and richness of ore. A recent visit to the mine has fixed in the writers mind that the Yankee Girl is the richest silver producer in America to-day, and that it will remain so until the progress of mining discovery and development gives it a new rival, which is possible if not probable. The product of the Yankee Girl up to the present is over $4,500,000. Had the property been operated for its fullest capacity of production, no one could doubt, after seeing the ore in sight, that its product could have been nearly or quite double that amount.

> *The Red Mountain Journal*
> Friday, 27 March 1891

In 1893, the silver crash forced the mines to close. Many of the residents left before the turn of the century.

By 1896, only 40 people lived in the town of Red Mountain and the post office had closed, according to *The Colorado State Directory* for that year:

RED MOUNTAIN

A mining camp in Ouray county in the Silverton Ry., 12 miles northeast of Silverton. Population 40. Altitude 11,300. Postoffice discontinued. Mail Guston.

Barrow, St James, justice pc
Collins, C M, jeweler
Henry, A E, contractor and builder
Kipp & Morris, saloon
Rumery, N E, meat market
Sheehan, Wm, supt Genesee Vanderbilt Mining Co
Thomas, G B, notary public
Waite, O S, lumber, timber

In 2000, Congress allocated $5 million so that the Trust for Public Lands could start buying lands in the Red Mountain Mining District. The National Trust listed these lands for Historic Preservation as among 11 of the nation's most endangered historic sites. Hopefully, this funding will help to preserve these remnants of Colorado's turbulent mining heritage.

Unidentified abandoned structure in Red Mountain, a now abandoned town named after the massive Red Mountain, known for the red stains streaking its sides.

SNEFFELS
Ouray County

Directions: Coming from the southwest edge of Ouray on U.S. Highway is a sign noting the direction to Camp Bird, which is Forest Route 853. Follow it. A four-wheel drive vehicle is recommended. After five miles you will be at Camp Bird. Drive carefully. The road gets very narrow, and you will be driving on a high shelf road with part of the mountain above hanging over the road. Make sure that no one is coming down the road toward you. At 5.9 miles, there will be a road on the left leading to Imogene Pass, disregard this turnoff and keep going straight. Almost immediately on the left across the river is the Wheel of Fortune Mine, and then on the right will be the Torpedo Eclipse Mine and Mill. After that at exactly 6 miles, will be Sneffels on the right side of the road.

In the late 1870's, rich silver strikes were made in the Mount Sneffels area. A camp was soon formed. The men named the camp Sneffels. Both the mountain (actually, a dormant volcano) and the camp were named after the icy mountain crater called Mount Sneffels in Jules Vern's classic tale, *A Journey To The Center Of The Earth*. The first known prospectors here were brought by two men named Quinn and Richardson. They helped to form the town. Not long after that, a prospector named George Wright came from Silverton, and staked his claim, calling it the Wheel of Fortune.

Sneffels was considered a silver town, but miners started to discover gold when further developing their claims. The deeper they dug, the more gold they found.

By 1877, many good mines had been discovered, and the Sneffels Mining District was created. Area mines included the Atlas, the Hidden Treasure, the Governor, the Humbolt, the Revenue, the Ruby Trust, the Senator, the Virginius, and the Yankee Boy mines. These were very rich properties that were discovered in 1877, with some of the best ore assaying at $40,000 to the ton.

The population hit 2,000. Even in winter, they came, braving the frozen roads and living in cabins that would be completely buried in snow. Avalanches were common.

By 1881, Sneffels had grown into a boomtown. Over 3,000 miners were now working the district.

The neighboring town of Ouray seven miles northeast of Sneffels and down in the valley was now a booming little town.

In 1882, 100 people lived in Mount Sneffels, according to *The Colorado Business Directory*:

MOUNT SNEFFLES
Mining Camp in Ouray County, 7 miles southwest of Ouray and 20 northwest of Silverton.

Mt. Sneffles M. & R. Co.
Porter & Brown, gen'l mdse.
Star S. & M. Co.

In 1884, J. A. Thatcher and M. D. Thatcher, two brothers from Ouray who owned the Thatcher Brothers Miners and Merchants Bank, helped finance the Revenue Tunnel. They invested $600,000 into the property. The Revenue Tunnel was built under the Virginius Mine. The two were essentially one great mine. The Revenue Tunnel was 3,000 feet beneath the Virginius Mine, extending it for three miles. This helped to drain and ventilate the Virginius Mine, which had been built seven years earlier. The Revenue Tunnel was a financial success and produced high-grade ore for 40 years.

The road from Ouray is a partial ledge road: steep, narrow, and dangerous. They say from 1881 to 1919, Sneffels' boom years, that the road was in constant use by wagons, freighters, coaches, miners, people on horseback, and long trains of burros, loaded with ore. All of the area mines had boardinghouses and cabins scattered around.

Between 1881 and 1919, the mines in Sneffels yielded $27 million.

The Colorado Business Directory for 1896 listed the following for Sneffels:

SNEFFLES
A mining camp in Ouray county, 7 miles southwest of Ouray, the nearest railway point. Altitude 10,300. Population 350.

Porter & Co, Geo R, gen'l mdse

Sneffels was built before Ouray and Camp Bird, and it produced almost as much ore. After production slowed down, the miners moved to Ouray, abandoning Sneffels. Today, Sneffels is nothing but scattered wood on the ground, pieces of metal, and shards of weathered glass. On the left side of the road across the river is the Ruby Trust Mine.

The narrow, rutted ledge road to Sneffels. At some points, the road shrinks to less than six feet across.

The Eclipse Mine, once a major producer, lies abandoned in Sneffels.

ALMA
Park County

Directions: Alma is six miles northwest of Fairplay on State Highway 9. With an elevation of 10,355 feet, Alma has the highest elevation of any active town in the United States.

Alma is no longer a boomtown, but it has been reborn as a new type of mining town and is far from deserted.

In the beginning, Alma was a travelers' stop and a junction house. A better route was developed that led to Buckskin Joe using the Fairplay-Montgomery Road. This helped Alma because people had to pass right through the town. Alma Station is located two miles south of town. When Buckskin Joe's mine stopped producing, many of the miners came to Alma. In 1873, the post office was established. Alma was a popular name in the early 1800's. At least four women named Alma were living in town when the name was picked.

In the early 1880's, Alma was a desirable place to live, with a school, stores, hotels, boarding house, its own newspaper, and its own smelter plant.

In the mid 1860's a gang called the Espinosa Brothers, nicknamed "the bloody Espinosas," murdered 32 people, including six residents of Alma. People had different stories of why they did this. The most often told rationale was that they saw an image of the Virgin Mary, and she told the brothers that they had to kill all the white people. Some say that some white men kicked the brothers off their property. Another story is that some white men killed their mother and father. But, perhaps they did it without any cause.

Central and Southern Colorado was terrorized by these men. A big bounty was put on the Espinosas' heads. A posse from California Gulch claimed to have killed one brother near Espinosa Peak.

Eventually a mountain man by the name of Tom Tobin, along with a posse, found the brothers and some members of their gang at the border of southern Colorado. Tobin and the posse killed them. As proof that the job was done, an Espinosa head was brought to Fort Garland.

In early times before Alma was here, an Indian village was very close to the Alma site. Fremont and Zebulon Pike were only two of the many explorers who reportedly had camped in Alma.

In the early 1860's, miners discovered gold above Buckskin Gulch. In 1870 silver was found, although it was hard to develop due to the altitude, inaccessibility, and poor smelting methods.

By 1873, Alma was a transportation center and the supply center for the district.

In 1880, Alma turned into a boomtown with a population that reached nearly 1,000 at its peak.

To date, the mines in Alma have produced over $50 million in gold, silver, copper, lead, and zinc.

The Sweet Home Mine was developed in 1872. In those days, the mine was mined for argentiferous galena. Nearly a century later, in 1965, the miners found a rich vein of strikingly beautiful deep red crystals of gem quality rhodocrosite. The Sweet Home Mine is now known for producing some of the most highly prized rhodocrosite in the world. Denver's Natural History Museum has an excellent display of these rare crystals. To see the mine, take Buckskin Gulch Road (Forest Service Road 416) leaving Alma. Go west to Buckskin Joe. On the way, on the high cliffs to the south, is the Paris Mine. If you look closely, you may see an old aerial tramway. From the Paris Mill, continue on the same winding road for another mile and a half. Buckskin Creek is the right of the road. This will take you to the property of the Sweet Home Mine.

If you look to the west from Buckskin Joe, you will find a gold arrastre built by the Spaniards in the early 1800's.

BUCKSKIN JOE
Park County

Directions: Buckskin Joe is two miles west of Alma.

The camp was named after Joseph Higgenbottom, nicknamed Buckskin Joe, for the leather he wore. Someone tried to name the town Laurette, but it did not stick, and Buckskin Joe it was. Buckskin Joe said he made the first strike here, but records indicate that this honor belongs to Hart Harris. Around 1860 Harris was deer hunting, and thinking he may have shot his quarry, went to see if there were any signs of blood. When Harris looked at the dirt where his bullet had struck, the earth sparkled in the sunlight. Harris had discovered a very rich lode of placer gold.

By midwinter 1864, when rich iron and copper ores were discovered in Buckskin Joe, *The Rocky Mountain News* described the find and the jubilant mood of the miners:

BUCKSKIN JOE.

We have some good news from Park County. The company that has been prospecting at so great expense on the Phillips lode have struck rich iron in fair quantities. A box full of the ore is on exhibition at Clark & Gruber's banking house in this city. A test shows it worth about one hundred dollars per ton, or eight hundred to the cord. Members of the company are jubilant, and each has his pockets loaded with specimens of the rock. It is iron and copper pyrites, with veins of white quartz, resembling that of California more nearly than any we have seen before seen in Colorado. Independent of gold, it must be very rich in copper, and there are in it occasional cubes of lead. The company is pushing work rapidly. The ore shown is from an incline on number four. They will begin crushing as soon as the weather moderates a little more. Very rich ore has also been struck on the Sublette. Specimens shown us at the banking house of Cass, Eaton & Co., assay five cents to the ounce, a yield so immense as to seem almost incredible. It is very fine grained iron pyrites, containing, apparently, a good deal of antimony. It will probably be found difficult to save all the gold it contains by common milling process. Good accounts also reach us from the Orphan Boy, Kitty Clyde, Honeycomb, and Union lodes. Some very fine specimens from the first named can also be seen at

Cass & Co.'s. The people in the country are in improved spirits, and look confidently forward to a big summer's work. A large amount of mining property has been bonded preparatory to the organization of one or more companies in the east.

The Rocky Mountain News
Wednesday, 27 January 1864

In 1880, The Carbonate Weekly Chronicle printed the adventures of an anonymous early settler who had been caught in a spring blizzard while crossing Mosquito Pass:

BYGONE DAYS.

Leaves From the Diary of an Old Settler.
A Terrible Trip Over the Mosquito Pass.
Lost in a Storm for Two Days.

Tales of the days of yore in this gulch possess a deep fascination for the average citizen and are read with interest by all. To-day we give an extract from the well-kept diary of a gentleman who, while he does not desire his name mentioned, is so thoroughly reliable and well known that the addition of his name would not add in any way to the interest of his narrative. The scene of his remarkable and perilous experience was at the head of what is now known as Bird's Eye gulch, just this side of the summit of Mosquito pass.

On May 19th, 1864, I started for California gulch from Denver in the "Buckskin Joe" coach. It was a beautiful morning when we left Denver, but toward the middle of the afternoon it commenced thundering and lightening off to the east of the road we were traveling. This was in the region of the great hail storm, which played havoc with the ranches of Plum Creek, and was the occasion of the great Cherry Creek flood, when Denver was swept away. It was fortunate for me that I was not in Denver that night, as my bedfellow, J. Bruce Haynes, was swept away in his bed, and his lifeless body was picked up two miles down the Platte. About twenty persons lost their lives, and the value of property destroyed was fully $200,000. It must have been a terrible sight to behold, and I consider myself extremely fortunate in having left Denver on the morning of this ill-fated day. Verily, "we know not what a day may bring forth." But to my story. We arrived in Buckskin Jo, which was then the metropolis of Park county, about 7 p.m. on Friday. Billy McClellan was our driver. Buckskin Jo had about two or three hundred people, H. A. W. [Horace] Tabor was postmaster and storekeeper, J. B. Stansell was managing the great Phillips lode, in which he was heavily interested, and Hart Harris was with him, being his partner. As soon as I arrived I began preparing for a trip across the range by the way of Mosquito pass. On the morning of Saturday, the 21st, myself, my partner, Mr. Frisbie, and a man named J. L. Dyer [Father Dyer], who carried the mail to California Gulch, started on what proved to be a perilous journey across the range. We proceeded very well until near the summit of the vast snow-field, when a heavy snow storm set in which completely blinded us, and made it next to impossible for us to tell in which three of us lost sight of our guide, the mail carrier, and as I turned around to speak to my companions about it, my feet suddenly slipped from under me, and I started at lightening speed down the precipitous side of the lofty snow-covered mountain. The clouds soon hid me from the sight of my companions, and for aught they knew, I was dashed to pieces upon the rocks below. But a kind Providence ruled it otherwise. I found by utmost efforts unavailing to arrest my downward career, until I had slid near one thousand feet, when a rather soft place in the snow and some loose rocks somewhat checked my speed, and I succeeded in stopping myself when about twenty feet from the brink of a rocky precipice, from which a sheer fall of two hundred feet would have certainly killed me. Though somewhat bruised I was not materially injured. Just then upon getting my breath I heard a pistol shot from my companions, who took this method of finding out if I were still alive. I shouted as loud as I could, and by frequent cries directed them to the spot where I was clinging to the rocks. It was an hour, and seemed an age to me, before I had the satisfaction of seeing my friends emerge from the clouds above me, crawling on their hands and knees. They assisted me and after hard work we succeeded in picking our way around the precipice and down into the gulch below. When we reached it we found that we were far to the right of where we wanted to go. After wandering around the remainder of the day, we built a large fire, made our beds, (each having a pair of blankets,) and ate our supper, which consisted of four Boston crackers for three men. In the morning we started again, breakfastless, upon our journey, and about three o'clock, p. m., arrived in California Gulch, tired, worn out and just a little hungry. The mail carrier, Dyer, had arrived safely at the gulch and reported there that he had left us on top of the range, getting along well, and presumed we would be in soon. As time passed and we did not come the citizens became alarmed, and a party started out in search of us. Finding no trace of us they returned and reported that we must have perished in a snow slide. The report soon got to Denver and Central City and caused quite a commotion among our friends. Gen. Sam Browne and others offered a reward for the recovery of our bodies, and the Central Register published a very fine obituary over a column long, with the heading "Perished in the Snow." We soon allayed the sorrow of our friends by writing to them, and in the language of the immortal Webster, "we still live" to tell the story.

The Carbonate Weekly Chronicle
14 February 1880

In the middle 1860's, the mines in Buckskin Joe stopped producing. The two remaining buildings, a dancehall and Tabor's store, were moved, restored, and set up near the Royal Gorge as part of a small amusement park, the Buckskin Joe Amusement Park. By 2000, the struggling amusement park went up for auction.

There is nothing left here except the site and the cemetery. The area is now ranchland.

COMO
Park County

Directions: To reach Como, take Highway 285 ten miles north to County Road 33 and then about another half mile up County Road 53.

Como is not completely abandoned. A few people live here and there in the community.

In Colorado's gold rush days, people knew Como for its roadhouse and then as a railroad stop in 1879.

Some coal mining was done here. There were a lot of Italian miners, and Como Lake next to the town was named after Italy's famed Lake Como.

In 1893 an area mine, the King Cole, exploded, killing 16 men.

Como's population peaked at 500. Thousands of travelers came through here on the way over Mosquito Pass, traveling to Leadville to seek their fortunes. This was a lively town with a hotel, saloons, a restaurant, stores, brothels, cabins, and an actual tent city with around 100 tents.

Como boasted a huge canvas tent with a 25-foot bar, gambling rooms, and a polished wooden dance floor.

As mining activity and traveling decreased gradually over the years, people simply did not go through Como anymore. In 1896 and again in 1909 fires raced through the town.

Then, when the railroad stopped coming into Como in 1937, the town nearly faded away.

FAIRPLAY (South Park City, Platte City, and Fairplay Diggings) Park County

Directions: Fairplay is about 20 miles south of Breckenridge on Highway 9. Or, if you are coming from Alma, go south on Highway 9 for five miles. It has an elevation of 9,896 feet. Today, it has a population of about 400.

The area was settled in 1859 after gold was discovered.

Fairplay was settled by a group of prospectors who had been turned away at Tarryall. The miners at Tarryall were stingy with their claims. The Tarryall miners had more claims than they could work, but they were greedy and refused to share the mineral potential of the area with newcomers. They chased away all the would-be prospectors coming to the area. The miners who had been chased away nicknamed Tarryall the Graball Camp, and they moved on until they found a beautiful spot along the South Platte River. This is where they settled. They found a little gold, built their mining district, made themselves log cabins, and then the prospectors got together to jointly choose a name for their town.

An ambitious prospector named Jim Reynolds spoke up and suggested that they call it Fairplay. Because this is where people would give other people a fair chance at trying to make a living, having a home, family, and a life.

Not long after Fairplay was settled, the Civil War started. Jim and John Reynolds [they were brothers] openly admitted that they were on the side of the South. They were taken to Denver and thrown in jail for having Southern sympathies. In jail, Jim and John Reynolds were imprisoned with Colonel McKee, Charley Harrison, and others just like them. Together, these men planned a successful escape. The men found their way to Texas. In Texas the Reynolds brothers joined McKee's band of rebels. In early 1864, Colonel McKee started to organize a gang with the intent of overrunning Colorado.

Colonel McKee gave the men a pass to get through the state line. Twenty-two men accompanied the Reynolds brothers north to the Santa Fe Trail. There, they robbed a big wagon train. They took $40,000 in cash and approximately $6,000 in draft notes. When the men got away, they started to argue over the money. Jim Reynolds threw it on the ground and said that they needed it to buy guns for the new recruits. They were going to try to recruit the miners who had come from the South. After hearing this, many men simply left.

Jim and John Reynolds, Jake Stowe, Owen Singleterry, John Bobbitt, Tom Holliman, Jack Robinson, John Andrews, and Tom Knight were members of the Reynolds gang. Although they robbed many people, they never killed anyone.

The Reynolds brothers and their eight remaining companions proceeded to Spanish Peaks and hid their money there. Riding to the Arkansas River, they crossed at South Park and rode into California Gulch. California Gulch was not doing well enough to hold their interest. The men rode back to South Park and robbed a stagecoach in Buckskin Joe. They took $10,000 in gold dust. After this, they robbed the stagecoach station, masquerading as a splinter group of the Confederate army up from Texas. With all of the Indian uprisings and attacks, the Reynolds gang thought that they could get away with this. The threat of a marauding Confederate Army in Colorado threw everyone into a panic. Military troops were forming and being trained. Denver troops guarded all of the stagecoaches.

The Reynolds gang had no clue of the scare they started in Colorado. They continued their crime spree, robbing travelers, stagecoaches, and ranchers. Oddly, they were very polite to their victims. One victim warned the gang that they were being tracked by a posse and that the pose would catch up with them anytime. After that, the Reynolds gang rode their horses into the deep woods, leaving the area and riding down the South Platte to Geneva Creek. There, in a cluster of pine trees, the rebels made camp for the night. Jim and John Reynolds went up Geneva Creek to bury some currency and gold. That night, Jim was dividing a bag of gold dust among the band, when out of the shadows a gunshot was heard. Jim Reynolds dropped to the ground, wounded.

A posse from Boulder County had accidentally ridden into the Reynolds gang's camp, triggering a gunfight. One of the gang was killed, one was captured alive, and the rest of the gang ran, not even wasting time to mount their horses.

The man killed was Owen Singleterry. The posse cut off Singleterry's head and took it, with their prisoner, to Fairplay. In Fairplay, the head was put on top of a pole and displayed publicly for a few days. Then, they Singleterry's head in a jar and preserved it.

The posse finally chased all of the gang and captured nearly everyone.

The six captives were taken to Denver and then sent to Colonel Chivington and his Third Cavalry, who were stationed to the south of Fort Lyon. After a few days the men were manacled, blindfolded, and shot. When the prisoners' bodies fell to the ground, the cavalry took the shackles off the men, leaving the unburied bodies to be eaten by coyotes. The cavalry was in a hurry. They had orders to march to an uprising at Sand Creek.

One of the Reynolds gang survived the gunfire. He was seriously injured, but he was able to crawl to a deserted cabin. After recovering from his injuries, he went to Denver to find his friends, Stowe and John Reynolds, the only other survivors. At Santa Fe he found them, then went to Spanish Peaks and retrieved some buried currency and gold. After retrieving their loot, the men needed fresh horses, and while stealing horses, the two bandits were killed, but John Reynolds escaped.

John Reynolds met a man named Brown, and the two men traveled and robbed together. Eventually, John Reynolds was shot from his horse. Brown tried to comfort and help his dying companion. John told him where he had buried the Buckskin Joe coach booty, and there was $40,000 in currency. They had wrapped the money inside oilcloth and with gold packed in three cans, and then buried their loot, Reynolds said. "You go up Geneva Creek. When you get to the head [top] of the gulch, go right and walk alongside the mountain a little while. At Dee Creek, stop, because just above the head of it at timberline in a prospect hole, we put our loot in it. Then we stuffed it up with stones. Ten steps below it, we stuck a knife in a pine tree, probably four feet high."

John Reynolds tried to sketch a map before his death. Brown buried Reynolds, but he was unable to find the treasure. To this day, no one has found it. In 1897 General Cook reported that the treasure was still hidden in the mountains.

There are many different stories surrounding where the treasure was hidden and how much money is still buried.

In one story, the loot is buried in a cave by Shafer's Crossing. A different story reports that it is buried a few feet out from from Webster's Pass Road, right where it crosses Elk Creek. Another story says the treasure was buried in what looks like a grave.

Vernon L. Crow, a respected old man who had lived in the Handcart Gulch area all his life, spent much of his life searching for the stolen loot. When he was young, the nephew of a member of the Reynolds' gang was a hired hand at Vernon Crow's parents' ranch. This hired hand searched for the treasure, and he told Crow the story that had been passed down to him. Crow found a tree pierced by a knife, but he never found the treasure. The creek has a lot of iron in it, so metal detectors do not work. Crow thought the loot was hidden up Handcart Gulch. A trail starts from the ghost town site of Hall Valley, about 10 miles from Grant. From beneath the ruins of the Hall Valley Smelter, Handcart Creek flows into the South Platte River. Half of a mile or so upstream was a very old stonewall, which resembled a rock ledge because it was so old and skillfully constructed. Going up a trail from this sign was an aspen grove with a little hidden park-like clearing. In the center of the clearing, a stone formation formed a line, pointing to remnants of a lean-to or shed. Higher up, the slope leveled off into a square area, about a third of an acre in size. On the right was a steep drop into Handcart Gulch that turned into Handcart Creek. Here, there were traces of an old trench that had once been a temporary canal. There was a clearing with some trees next to it. Cut timbers were piled nearby. A little over a mile or so upstream the creek from here was a grave. Crow dug up this grave, and he found a skeleton, still wearing boots, with a bullet in his skull. A nearby tree had scars on the bark. This is all Crow said he could find of the stolen loot.

HAMILTON
Park County

Directions: Hamilton is directly across the creek from Tarryall. It is two miles north of Como on County Road 35 where it crosses Tarryall Creek.

Hamilton was founded in 1859 by prospectors who had been turned away from the town of Tarryall. The original Tarryall miners passed a law saying that their mining claims could be 150 feet long, at a time when no one else had claims over 100 feet long. This left no room for any other miners. The miners who were turned away crossed the creek and started their own town. Reportedly, they wanted nothing to do with Tarryall after such bad treatment.

Hamilton, named for Earl Hamilton, one of the original founders, welcomed everyone. The town boomed. In a year, the population had blown up to 6,000, with saloons, a theater, gambling houses, and a newspaper, *The Miners' Record*. In 1861 Governor Gilpin visited Hamilton. Arapahoe and Ute Indians camped nearby and lived peaceably with the white miners.

The Colorado State Directory for 1878 listed the following on Hamilton:

HAMILTON
Mining camp in Park County, 12 miles north of Fairplay. Population 100.

Attorney: W. Ballinger.
Hotels: T. Dunbar, H. H. Lienthal.
General Merchandise: A. Blandin.
Miners: Peabody & Crozier, Curtis & Barrett, S. Taylor, W. Leobelt.

In 1880 the gold in this area started to become scarce, and within a few years people left. By the mid 1880's Hamilton was a ghost town. In the early 1900's, dredging operations were started. There is no sign that a town ever existed here, just tailings piles.

By 2000, there were still had about ten inhabited houses and the original schoolhouse in Tarryall.

HORSESHOE (Doran, East Leadville)
Park County

Directions: Leaving Fairplay on US Highway 285, turn right on Four Mile Road. Horseshoe is eight miles west.

Horseshoe was established in 1879, after silver was discovered here. The town was named after a natural horseshoe-shaped mark on the side of Horseshoe Mountain. Horseshoe thrived for 15 years. In the summers the population peaked at 800. There were hotels, saloons, a few stores, smelters, and mills.

In 1893, the silver crash destroyed the town. Some residents remained and attempted to revive the town. In 1896 when a railroad went by the town, a few residents tried to breathe new life into their dying town, but their effort failed. In 1902, yet another attempt was made. Someone even tried to rename it Doran, but the town was deserted by 1907. A final, half-hearted revival in 1920 failed.

The last two original buildings were restored and moved to South Park City in Fairplay.

JEFFERSON
Park County

Directions: Jefferson is located on U.S. Highway 285, about 16 miles northeast of Fairplay. Jefferson is 81 miles southwest of Denver.

Jefferson sprang up as a mining camp around 1865; the exact date is unknown. It is said that Jefferson, named for Thomas Jefferson, was started by miners who worked at nearby camps. They had panned some gold, and then it ran out. The miners had hoped to make this town the capitol when this territory was formed into a state. In 1879, the railroad came through Jefferson on route between Denver and Fairplay. It helped to revive the town since the gold dried out here.

The Colorado Business Directory for 1882 listed Jefferson's population at 50. Jefferson was a tent city and a railroad station for the D. & S. P. Railroad.

The town is now deserted and empty, except for the train station, which had been restored. The train tracks were ripped up sometime around 1930.

LEAVICK
Park County

Directions: On U.S. Highway 285 leaving Fairplay, turn right on Four Mile Road. Leavick is 12 miles west from Fairplay.

Around 1873, the Last Chance Mine was founded up the gulch from Horseshoe. On one side of the mine was Mount Sherman, and on the other side was Mount Sheridan. In 1885, a mill was built below the mine. Miners, who had been camping out across from the mill, soon started a little town, which went unnamed until 1896. And then it was named after prospector Felix Leavick. For the past four years, he had been the owner of the Hill Top Mine, which was one of the better mines in the county. The mill for the Hill Top had a narrow gauge railway stop in front of it.

Leavick had a few businesses, some saloons, a dentist, barbershop, parlor house, a tramway system for the mines, and a mill.

The population peaked at 300, but the silver crash of 1893 wiped out most of the population. Mining continued until around 1930 or so, but the few miners left were mining zinc.

Vandals destroyed many of the old buildings. The Historical Society of South Park was able to save the last two original buildings, the Homestead House and the Hoffman Blacksmith Shop, and moved them to South Park City in Fairplay.

The Leavick site is easy to reach, but nothing is left.

LONDON JUNCTION and ALMA STATION
Park County

Directions: London Junction is on State Highway 9, two miles south of Alma.

London Junction is adjacent to Alma Station; the two communities are only about 150 feet apart.

The North London Mine was discovered in 1875. The North London Mine had two tunnels: the London and the Hard to Beat. To see what is left of the mill and some of its cabins, find Mosquito Gulch and follow the north fork west. It is seven miles from Alma.

The railroad was established here in 1882 and an aerial tramway was used to transport the ore from the mills to London Junction.

In 1883, the owner of the South and North London Mines had built its office and a large mill at London Junction. The mill had 20 stamps to crush ore.

The boarding houses for North London are on the north side of London Mountain. To get there from south edge of Alma, on State Highway 9, you turn left and get on State Highway 300. After three miles, the road will fork. Go right and there will be what is left of the boarding houses. This should be up Mosquito Gulch, with Mosquito Pass not far.

To find the South London Mine, take State Highway 9 to Alma. On the South edge of Alma is a sign for Mosquito Pass, directing you to a dirt road to the left. Go left. When the road forks, turn left. Off this road you will find South London Mining and its buildings and cabins where the miners lived.

These London Mines produced more than $8 million in gold, silver, and other minerals.

Many miners lived in the area of London Junction and at Alma Station. They did a lot of placer mining.

The area still has a few residents.

PARK CITY (Mosquito)
Park County

Directions: Coming from Alma, which is located on State Highway 9, go directly west into Mosquito Gulch. Take this road for three miles, and you will be at Park City.

The small camp of Park City started to coalesce in 1861. The camp stayed small for years, but in 1879 Mosquito Pass began to grow. Park City was at the base of Mosquito Pass, on the main route to Leadville. A little bit of mining was done here, but this area also had an important stagecoach stop. Thousands of miners passed through here.

In 1880, there was a saloon, the Park Hotel, a boarding house, a store, and a brothel. *The Colorado Business Directory* for 1882 listed the population at 200.

In 1882, the railroad arrived. Mosquito Pass was such a deadly pass that hundreds of people who did not have coach fare to Leadville died of exposure trying to walk the pass. This is how the road earned its grisly nickname, "The Highway of Frozen Death."

Park City started to die. The Halfway House and the assay office were restored and moved to South Park City in Fairplay.

Nothing is left of Park City today.

TARRYALL
Park County

Directions: Tarryall is on County Road 33. Coming from Como, go two miles north. The town was across the creek from Hamilton.

Gold chunks the size of watermelon seeds were found here in 1859, but the city was not platted until 1861. Prospectors had arrived and set up camp in this area, only to find the ruins of a previous camp, cabins and buildings that had been at least ten years old. It is only speculation, but Indians may have killed all the miners of this early camp.

John Fremont reported that a man named "Parson" Bill Williams discovered gold in 1849. Could this have been the place? In his journal, Zebulon Pike documents a gold discovery in 1805 just west of here by James Purcell. Maybe this was that place.

In 1861, the population peaked at 300 people. Tarryall was the county seat for Park County for a little while. There were a lot of saloons here. It was said that everyone in town drank whiskey.

Sometimes nicknamed "Tarryall Diggings," this town had its own private mint, owned by John Parsons. His mint stamped out $2.50 and $5.00 gold pieces. Tarryall produced $2 million in gold before the mines dried up.

By 1875, the town was completely deserted. In the early 1900's, dredging operations buried most of the town under rocks and gravel. All that remains is a white, one-story schoolhouse, surrounded by several old cabins.

Today, only an old white wood schoolhouse looks over the remains of once boisterous Tarryall. The saloons are all gone. By 1875, the town was completely deserted.

WEBSTER
Park County

Directions: Webster is 69 miles southwest of Denver and three miles west of Grant on U.S. Highway 285.

This little camp was basically a railroad and wagon supply station at the foot of Webster Pass. This important pass was built in 1878 by William Emerson Webster, for whom the town was named, and Montezuma Silver Mining Company. Supplies were brought through here for the mining districts. Webster Pass has snow on it almost all year long.

The Daily Journal from Breckenridge reported the following news on the toll road between Webster and Breckenridge:

MINING NOTES.
The Route to Webster.

On Tuesday of last week the toll road across Hand-Cart pass, from Webster to this place, was opened for wagon travel, the distance is fifteen miles and affords a short route from the mining towns in the eastern portion of the county to the Denver & South Park railroad. Freight rates are $1 per hundred to Montezuma. Frank S. Taylor has taken the contract for tri-weekly mail service, and has established a passenger and express line, leaving Webster Mondays, Wednesdays, and Fridays, and alternate days from Montezuma. At Webster are hotel and telegraph accommodations for all having business in that vicinity. A livery is connected with the Webster house.

The Daily Journal
Wednesday, 29 June 1881

In 1882, 75 people were living in Webster, according to *The Colorado Business Directory* for 1882:

WEBSTER
Station on Denver & South Park railroad, 69 miles southwest of Denver, 5 from Hall Valley. Daily mails.

Champian & Tandey, forwarding and general merchandise.

Smith, James J., postmaster.
Smith & Croft, blacksmiths.
Webster House, Grow & Smith, proprs.

Webster was a rowdy, rough mining town. Two cemeteries were on the north side of town: one was for shooting victims, travelers who died passing through town, and transients; the other cemetery was for Webster's residents. The one hotel, Webster House, was very busy. For a dollar, you could sleep on the floor for the night. Sometimes you got a blanket, and sometimes, you did not.

As time went by, easier and better passes were constructed. When people stopped using Webster Pass, the little town lost its reason for being, and its residents drifted to other towns.

JASPER (Cornwall)
Rio Grande County

Directions: Jasper is seven miles east of Summitville on Forest Route 250.

When Jasper was settled in 1874, it was called Cornwall, after John Cornwall, the postmaster and one of the town's founders. Cornwall's Mountain is a few miles southwest of Jasper. When all of the rich strikes were made at nearby Summitville, prospectors flooded the area, searching for the next strike. Two of the best mines, the Perry Lode and the Miser, were discovered at the base of Cornwall's Mountain. The Sanger Mine was two miles above Jasper.

The owners said the mountain was filled with bird's-eye porphyry. The owners found as many investors as they could. Then they skipped town with all the money. The fraud was on such a grand scale and so blatant that everyone heard about it. After this, no one wanted to invest in Jasper, and the town could not get funding to develop the mines, properties, or the town.

In 1885, 200 people were living in Jasper, and the town boasted one stamp mill, according to *Crofutt's Grip-Sack Guide of Colorado*. In 1887 miners and prospectors in Jasper did not know the value of their ore. They shipped ten tons of their ore off to a smelter in Denver to be refined. A fire destroyed the smelter, along with all the townspeople's ore. This was the final death knell for the little town, and many of the residents moved away. If you walk past the inhabited summer cabins and along the creek, you will see what is left of the old town of Jasper.

The Colorado State Business Directory for 1882 listed the following on Cornwall:

CORNWALL

Mining camp, 38 miles west of Alamosa, Rio Grande County. Tri-weekly stage to Alamosa. 6 miles east of Summitville.

Cornwall Silver Mining Company.
Little Annie Mining Company.
Millington & Moody, mining contractors.
Millington, Wm. V., postmaster.
Swanson, F. W., general mdse. and saloon.

SUMMITVILLE
Rio Grande County

Directions: Summitville is 27 miles southwest of Del Norte on Forest Route 330. It is 308 miles from Denver.

In 1870, two brothers, William and James Wightman, along with others, discovered gold east of South Mountain, in what is now called Wightman Gulch, at an elevation of 11,300 feet. By 1871 when spring came, the area was filled with prospectors, but by the end of the summer everyone had left except for the Wightman brothers and two of their companions, Peterson and Johnson. In 1872, prospectors returned to seek their fortunes, but it was not until 1873 that the Little Annie was discovered. The best mine in Summitville, Little Annie held a seemingly inexhaustible vein of gold that was 17 feet wide. The ore ran between $80 and $2,000 a ton.

John Esmond was the original founder of the Little Annie, but he failed to file the paperwork to stake a claim. When Esmond returned to his diggings in the spring, he found that someone else had staked a claim, built a cabin, and had moved in. In response, John Esmond staked two claims in the area: the Esmond, later renamed the Aztec, and the Mayer.

Over the next few years, mining claims proliferated all over South Mountain, 2,500 all together, but only about ten were developed.

In the 1880's, Summitville became the top gold producing gold district in Colorado.

In 1882 *The Colorado Business Directory* reports the population of Summitville as 400. It also reports that Summitville was 27 miles southwest of Del Norte. The main transportation to town was stagecoach.

By 1885, the population had grown to 500 according to *Crofutt's Grip-Sack Guide of Colorado*. Telephone service had just been established

By 1886, 600 people lived in the town, and there were at least a dozen saloons. The local newspaper was *The Summitville Nugget*. The town was booming, and times were good for a few more years.

By the late 1890's, the town's rich veins of gold had been depleted, and Summitville was virtually abandoned.

The Colorado State Business Directory for 1896 listed the following:

SUMMITVILLE
Mining town in Rio Grande county, 28 miles southwest of Del Norte. Daily stages to Del Norte. Altitude 11,000. Population 50.

Clark, M J, books and stationery
Pangborn, Asa H, saloon
Weiss Bros, A J, genl mdse

Summitville was abandoned until the fall of 1934 when Summitville Consolidated Mining Corporation revived the town. With a considerable amount of money and owning over a hundred claims, the company built boardinghouses, 70 homes, a school, mess halls, a bathhouse, a recreation center, a post-office, and a new water system. The population reached 700, and the mine was producing $70,000 in gold a month. Summitville had evolved into a company town.

In more recent times, Summitville is known for the controversial case of the "Summitville Boulder." On 3 October 1975, a contractor for the Reynolds Mining Company, Bob Elithorpe, was surveying the gold resources around Summitville. He thought he saw the glint of gold in a roadside boulder outside of Summitville. Bob Elithorpe returned to the camp, and asked a company geologist to look at the rock and maybe help him put it in the back of his truck. When the geologist examined the boulder, he confirmed that it was gold. In his excitement, Elithorpe asked the company geologist if he could keep half as a finder's fee. Ten percent was the typical finder's fee for reporting the location of a mineral deposit, but this was finding a massive nugget, so Elithorpe thought the percentage should be higher. The geologist verbally agreed without giving the matter another thought. The geologist said that the boulder had been sitting on the side of the road fully exposed for 30 years. When testing revealed that the boulder contained 350 ounces of gold, the company rejected Elithorpe's claim. The geologist, they contended, did not have the legal right to make any decisions for the company. Elithorpe was not a prospector, he was a company-paid contractor working on company property, and therefore was not entitled to the traditional finder's fee. There was no legal precedent for Elithorpe to claim the Summitville Boulder. Reynolds Mining Company later offered Elithorpe a $21,000 finder's fee, which he accepted. The company later donated the boulder to the Denver Museum of Natural History where it is now on display.

Mining in Summitville continued sporadically until 1992, when the Summitville Consolidated Mining Corp. declared bankruptcy. The company had extracted about $180 million in gold between 1986 and 1992. It excavated a vast 572-acre pit, and then used cyanide to leach out gold particles from crushed rock. Each spring, these chemical residues of old mining processes wash downstream into the nearby Terrace Reservoir and the Alamosa River watershed.

Summitville is now abandoned, except for massive clean up efforts. In 1994, the EPA named the Summitville mine a Superfund site, epitomizing the cyanide and heavy metal pollution of Colorado's open-pit gold mining history. During the 1990's, the Environmental Protection Agency spent more than $150 million in its efforts to clean up the area. As of 2001, the effort was not complete.

BONANZA (Bonanza City)
Saguache County

Directions: Bonanza is 261 miles from Denver. From Buena Vista, take State Highway 285 south. About half a mile before you reach the little town of Villa Grove, turn right on County Road LL56. This road follows Kerber Creek through a valley dotted with cottonwoods until it reaches Bonanza.

In 1880 prospectors discovered veins of silver, lead, and magnesium in the area of upper Kerber Creek. The creek was named after Captain Charles Kerber from Fort Garland who passed through the area in 1865. Reportedly, a man was looking for some horses in Copper Gulch, when he found some rich float. He filed a claim, calling it the White Iron. When news of the gold claim spread, hopeful prospectors set up mining camps along Kerber Creek. Perhaps 4,000 people came to the new camp of Bonanza.

When one of the first strikes was made, a prospector shouted, "Boys, she's a Bonanza!" That is how Bonanza got its name. The residents originally called it Bonanza City, but later shortened the name to simply Bonanza. The town was located at the bottom of a narrow gulch, with high mountains on each side, with Kerber Creek, and Main Street running right through it. At one point the creek and the road cross each other.

Originally, five towns were here: Bonanza, Sedgewick, Kerber City, Exchequerville, and Bonita. Sedgewick was one mile southeast of Bonanza. Its population was 650, and it was the first town built in the area in July of 1880. There was a big brewery here and a bowling alley. Across the creek from Sedgewick was the little town of Kerber City. About a mile northwest of Bonanza was the town of Exchequerville. Exchequerville was located at the junction of two old dirt roads: one leading to the Exchequer Mine, then the Shirley Mine; the other leading to the Raleigh Twelve, a more recent mine. Southwest of Sedgewick was the town of Bonita, but there is nothing to mark the existence of this little long-gone boomtown.

As Bonanza grew, these small towns were absorbed and became neighborhoods of Bonanza. In 1881 Bonanza was referred to as a new Leadville. The local newspaper was *The Bonanza News*. There was a big business district; at its peak, Bonanza had seven dancehalls and 36 saloons. Saturday nights often turned into big drunken parties.

Anne Heister Flemming Ellis was only six years old when she moved to Bonanza with her family in the fall of 1882. Her autobiography, *The Life of an Ordinary Woman*, depicted her struggle for survival in Bonanza, and later in other mining towns. Her book was published in 1929. Anne Ellis died in 1938 and was buried at the Exchequer Cemetery northwest of Bonanza.

The Summit County Circular for 18 February 1882, reported mill runs from the Argo mine:

GENERAL MINING NOTES.
The Empress Josephine at Bonanza, Colorado is one of the representative mines in the Kerber Creek region, and is improving as developments are being made. The following mill runs from the three grades of ores, recently made at Argo, will show the richness of the mine. The run was made on one ton of each grade: First grade, $773.77: second $276.75: third grade, $133.75.

The Summit County Circular
18 February 1882

In 1880, former President Ulysses S. Grant toured the area. After seeing the Bonanza Mine, he unsuccessfully tried to convince the owners to sell the mine to him for $40,000. The owners of the Bonanza were charging curious people a dollar a head to just look down the mine hole. At Bonanza Peak, the Bonanza and the Exchequer were some of the first strikes in the area. After the Rawley was found in 1881, it sold almost immediately for $100,000. In Copper Gulch, the Empress Josephine produced $7 million in ore. Reportedly, the Empress Josephine was reopened briefly in 1947. On Mineral Hill, the Defiance and the Wheel of Fortune were good producers, but it was discovered that the claims overlapped. The mine owners were able to work things out without getting their mines tied up in litigation. The ore in the North Kerber Creek area ran in pockets and was plentiful, but low grade. Unfortunately, low-grade ore was expensive to refine. In 1882, the price of silver plunged. People were scared and started to leave. A businessman/ Indian scout from Del Norte, Mark Biedell unsuccessfully tried to build a mill to effectively process low-grade ore.

By 1882, the population dropped to 1,300. Within three years, the population had fallen to 500.

In 1896 *The Colorado Business Directory* listed the following on Bonanza:

BONANZA

Mining town in Saguache county. Daily stage to Villa Grove and Saguache. Distances- Saguache 36 miles, Villa Grove 16. Population 250: altitude 9,375.

Anderson, Chas, 2nd hand gds
Baldwin & Mackay, gen mdse
Empress Josephine Min`g Co.
Kenny, Jas, postmaster.
New York Smelting Co.
Smeltzer, Nelly, millinery
Story, S C, meat market
Timney & Kouts, mt mkt

The population of Bonanza continued to drop, and by about 1910 the town was abandoned. There are some summer cabins in the area, but no services are available.

Bonanza is so far from the main highway that the area is comparatively untouched. Good specimens of ores and minerals can still be found in the relatively intact tailings piles.

In 2000, the Federal Census reported seven residents

IRIS
Saguache County

Directions: The easiest of many routes to Iris is to leave Gunnison going east on Highway 50. After about five miles, turn right on County Road 42. Go south for about eight miles. At this point, you will reach County Road YY10, turn right and drive for about ¾ of a mile. Iris is on the right, right next to the road.

The town of Iris was established in 1894, named after all of the wild irises in the fields. Iris had a sister city also founded in 1894 called Chance exactly one mile northwest, just across the county line in Gunnison County. Iris was slightly bigger because it was closer to the mines on Mineral Hill. Together, the towns had a population of about 1,000. Iris had telephone service, mail delivery three times a week, and a main street with all of the usual businesses. Iris was on the edge of the Gunnison gold belt.

In 1896, investors had put over $100,000 into the mining property in the area, but this was not enough to save the area. The several mines in the area simply were not productive.

Iris was fighting for survival in 1896 when *The Colorado Business Directory* printed the following:

IRIS
Mining town in Sagauche county, 14 miles southeast of Gunnison nearest railway point.

Buchanan, Jas, saloon
Eichelbergher, E C, mining and milling
Fitzpatrick, F J, saloon
Fetcher, E C, justice peace
Grant Bros, blacksmiths
Hudler, Clark, gen mdse
Kearney, S D, carpenter
Kirwin, J J, carpenter
McDonough Bros, livery
Mason, D W, mine supt
Notmore, CW, millwright
Piper, J P, groceries
Powell, Oscar, painter

Welch Bros, saw mill
Westfall, W J, teacher
Zimmerman, F, dairy

The towns of Iris and Chance faded away after the people left in 1900. In 1901-1902, a mining company tried to revive the town, but was unsuccessful.

LIBERTY
Saguache County

Directions: Liberty is located at the north edge of the Great Sand Dunes on the south-eastern edge of the Baca Grant #4, which is now state land. Liberty is about 15 miles to the south of Crestone. It is almost impossible to find Liberty. Coming from Mineral Hot Springs, take State Highway 17 south for about 26 miles to County Road D. Turn left, going east and follow the road for four miles until it ends in a jeep trail. Follow the trail. After the first mile, the road forks; stay right. After another mile the road forks again, stay left. Stay straight the rest of the way, and ignore any turnoffs. At the last fork in the road stay right and this will take you to Liberty. If you go right, this will take you to the site of Duncan. Liberty is 13 miles from the highway.

In 1882, the King of Spain gave one of his subjects a land grant of 100,000 acres in Saguache County. This is called the Luis Maria Baca Grant #4. This grant was later turned into a ranch. In the 1870's, when mining camps sprang up all over Colorado, so did camps in the grant. There were Cottonwood, Duncan, Julia City, Lucky, Spanish, and Teton. Cottonwood and Duncan had a combined population of 2,000.

In July 1900, the new owners of the grant, The San Luis Valley Land and Mining Company, had a U.S. Marshal serve summons to evict all of the miners in the town within the boundaries of the grant. There was nothing the miners could do; they left their homes, their mines, a way of life they had worked so hard to build. Most of the residents of Duncan moved what they could of their buildings across the eastern border of the grant property, forming a town they named Liberty. They named it Liberty in defiance because they were free of the constraints of the grant. The men tried to mine in the Liberty area, but without success. By 1910, everyone had left except for an eccentric old hermit who drank heavily and reportedly built mounds of animal bones in his front yard.

Nothing is left of Liberty except the spirit of those long-dead evicted miners.

A century later, in 2000, federal agencies have started a program to buy the 100,000-acre Baca Ranch to extend the existing Great Sand Dunes National Park.

ANIMAS FORKS
San Juan County

Directions: From Silverton take State Highway 110 northeast for 13 miles. You will drive by Howardsville, Middleton, and Eureka. State Highway 110 will take you right into Animas Forks.

In 1875 prospectors made their first gold strikes here. It was isolated and close to the timberline at an elevation of 11,500 feet. Prospectors came to the area from all over, and a city of tents sprang up at the junction of the east and west forks of the Animas River. A mill was built to process the ore that was shipped in from Mineral Points Red Cloud Mine.

A town was not formed until 1877, two years after the first gold strikes. The town founders advertised free property and free help to build a home, for anyone who might want to settle in this new town of prosperity called Animas Forks. And the people came. In those days a common route to Animas Forks was to come from Lake City on County Road 30 to Whitecross and then to Burrows Park, going over Cinnamon Pass to Animas Forks. Another common route from Lake City was to follow Henson Creek on County Road 4, southwest over Engineer Pass, then south to Mineral Point, and then south again for another few miles to Animas Forks.

These two routes are old trails, now considered rough four-wheel drive trails.

By 1881 the town had a main street lined with many businesses. That same year Animas Forks had a phone line installed. The phone line had been run from Lake City 20 miles away, crossing the Continental Divide at Engineer Pass. That Christmas, townspeople (nicknamed "Forkites") celebrated together in a big hall. A large fresh cut pine tree was decorated and surrounded by gifts for all of the children. The women had prepared turkeys, chickens, mutton, oysters, eggs, and eggnog.

In 1882, *The Colorado State Business Directory* listed the population at 250.

The local newspaper, the *Animas Forks Pioneer*, covered the local Independence Day celebration:

Animas Forks Pioneer
Geo. N. Raymond, Publisher
Marc. G. Perkins, Editor
Published at a Higher Altitude than any other Newspaper in the World.
Terms:-$3.00 a year; $1.50 for six months;
Devoted to the mining and local interests of Animas Forks, Mineral Point, Eureka, Burrows Park, Engineer Mountain District, and surrounding localities.

TOWN OFFICERS
Mayor- John R. Hunt
Recorder-Albert Dyes
Treasurer- Fred G. Heimbolt.
Trustees-S. W. Raymond, Ed. M. Brown, Fred G. Coombs and Charles E. Kimber.
Marshal-Robert Wiedemann.
Fire Warden- Fred Coombs

Fourth of July at Animas Forks.
The Fourth was celebrated at the Forks by exploding giant cartridges, displaying of white shirts, linen collars, etc., and in the evening there was a dance at Steins Hall, which was a successful affair. Among those present we noticed, Mr. and Mrs. C. E. Kimber, Mr. and Mrs. Frank Stein, Mr. and Mrs. Foell, Mr. and Mrs. Capt. McFarlane, Mr. and Mrs. F. G. Heimbolt, Mr. and Mrs. Duncan, Mr. and Mrs. Fred Aderholdt, Mr. and Mrs. Wm. Stein, Mr. and Mrs. Short, Mr. and Mrs. McFarlane, Mrs. Eckart, and the Misses Stanley. Messers. Tom Kane, Dick Trezona, J. H. Crist, and Ed Beatty, of Mineral Point: S. W. Raymond, C. H. Raymond, T. M. Walters, J. L. Stanley, Henry Ludwig, Chas. Carlstrom, J. A. Wright, Frank Thaler, G. D. Lockwood, E. P. Howard, Fred Harris, Z. McConnell. After the dance a fine supper was partaken of, which was gotten up by Mrs. Foell and the Misses Stanley, and was pronounced an elegant repast by all the partakers thereof.

Animas Forks Pioneer
Saturday, 8 July 1882

Interestingly, the *Animas Forks Pioneer* also printed an early feminist complaint in the form of a protest poem on Saturday, 23 September 1882, in addition to local gossip and the society notes.

A WOMAN'S PROTEST

As a woman standing all alone,
I humbly hope to shine,
I'm tired of the twaddle
Of the Oak and Ivy vine,
I've seen too many instances
Of natures law declining,
The vine did all the supporting
While the oak did all the twining.
Before I'd marry a man
Who'd place himself upon the throne,
And claim me for his "better half,"
Allegiance blind and mute,
I'd marry the nearest ape
And wait for him to evolute.

---- Queen Bee.

Or marry an Indian Chief
And be a belle Pi-ute?

Animas Forks Pioneer
Saturday, 23 September 1882

A number of freight teams came in this week loaded with supplies for the winter.

Great activity is being displayed by Forkites these days by getting in their winter fuel.

Fred Coombs has put a coat of paint on his house and improved the appearance of it very much.

McEntee Bro's of Mineral Point, are fixing up their store and getting in a large stock of goods for winter.

A hard snowstorm started in Sunday and lasted 48 hours. The snow is all gone and we are having charming weather.

PERSONALS

Dr. Murray made a visit to Silverton Wednesday.

Prof. Taylor and wife left for England last Tuesday.

Capt. A. W. Burrows returned Thursday from Del Norte.

Mr. and Mrs. E. B. Greenleaf, have returned from Wagon Wheel Gap.

Roberts, of Lake City, came over Monday to look after insurance matters.

Geo. J. Richards, of Lake City shot through town like a comet Wednesday.

J. L. Stanley went over to Lake City last Tuesday to purchase supplies for winter.

Judge Hollingsworth, of Silverton, came up Tuesday to see how Animas Forks looked.

Mrs. James Reynolds was quite sick the first part of the week, but is now convalescing.

Miss Alice Lord departed for Del Norte last Tuesday, leaving another gap in our social circle.

Mrs. P. McEnany came over from Lake City this week and is stopping at Mineral Point.

Wm. Young, one of the principal stockholders of the Red Cloud company came in Tuesday.

Charles D. Adams, our next county commissioner, came down from Mineral Point yesterday.

Abe Wright and A. L. Hall, went over to Lake City Thursday and took in the Pitkin Guards' Ball.

Animas Forks Pioneer
Saturday, 23 September 1882

Winters were brutal, and snow fell from October through April, often reaching a depth of 25 feet. From December to April, no person or building was safe from avalanches. Every winter brought its casualties. The Victorian homes were very well built, with quality lumber, quality shingles, gables, and beautiful bay windows. It was an attractive town, built on the top slanted edges of a gulch. The only ugly building in town was a small, two-cell jail that resembled a huge wooden box. Remarkably, in the mid 1880's when the population peaked at over 1,000, the jail was barely used. As the town grew, Otto Mears built a railroad, the Silverton-Northern, which ran from Silverton to Animas Forks.

A huge two-story Victorian home with a large bay window overlooking the Animas River dominates the remaining structures in town. Some sources report that Thomas Walsh owned and lived here before coming into his millions with the Camp Bird Mine in Ouray. His daughter Evalyn Walsh Mclean bought the Hope Diamond. She wrote an autobiography, *Father Struck It Rich*. Reportedly, she returned to Animas Forks to write her book. A conflicting source says that there is no proof that the Walshes ever lived in Animas Forks, the house had been built by William Duncan, and he and his family had been the only ones to live there.

The Colorado State Business Directory for 1878 listed the following for Animas Forks:

ANIMAS FORKS
Mining camp in San Juan County, 13 miles above Silverton. Population 300.

Postmaster- Brown, E. M.
Assayer. Brown, E. S.
Blacksmith. Ferguson, Wm.
General Merchandise. Lueslay, R. C.

Miners Supplies. Stein Brothers.
Hotel. Flag Staff Hotel.
Concentration Works. Greenleaf, E. B., Supt.
Restaurant. Koestle, Jacob.
Saw Mill. Brown & Camp.
Meat Market. Pierson, C. & Co., Pierson Bros. & Scribner.

On 20 January 1883, the *Animas Forks Pioneer* printed the following advertisements:

TIP TOP RESTAURANT
MRS. EKKARD, Proprietress.
Main Street, second door to the north of Stein Bros` General Store. This restaurant will be opened to the public on Monday next August 21st, by Mrs. Ekkard, long and favorably known as a first class coo [cook]. Everything new and complete and the table supplied with the best the market affords.

A. Wright. Chas. Carlstrom
WRIGHT & CARLSTROM
Dealers in
Fine Whiskeys
Imported and Domestic Cigars.
We carry a large stock of the best brands of all liquors, such as Brandies, Wines, Champagnes, Ales, Porters, Etc.
When you desire a Good Drink or a fresh Cigar, come and try us.
We also have in our room a fine Billiard Table and in every way we shall endeavor to make our place

A POPULAR GENTLEMAN'S RESORT.

The *Animas Forks Pioneer* printed the following article Saturday, 3 February 1883:

. . . It seems he [an unidentified, prospector, one of many who died unknown and unnamed in the quest for gold] did not appreciate Mr. Carey's warning, and started for the Forks on

Tuesday, returning in the late afternoon when snowslides are apt to run. The slide, though small, came down from the right hand side of the gulch, through a small draw on the mountainside, and was about 40 feet wide between the trees. It is evident that the slide had but little force when it reached him, for it ran only ten feet from where he was found and supposed to have been caught. When found his snowshoes were within four feet of him and his snowshoe pole laying by his side. It seems almost impossible for a slide so small and slow to catch a person in position, unless the strong wind and blinding storm prevented his hearing and seeing it. It is generally believed that the slide struck his feet tripping him up throwing him head first into the moving snow. When found his feet were projecting from the snow and his body and head were covered under about three feet. He was instantly killed by all appearances, as his face was badly mashed in and his body pressed out of shape. His body was removed from the slide on Thursday by Mr. Carey, McCloud and Machmer, and placed in one of the buildings near the Eclipse smelter where it will remain for several days, when it will be properly inturned. It is not known whether the poor fellow has any friends in this section of the country or not, but nevertheless his remains will be placed where his relatives may get them when the spring comes. Thus one poor soul who has dared to face the hardships of this rough and wild country passes into the unknown where snowslides and snowshoes do not haut [haunt?] the visions of their peaceful rest.

Animas Forks Pioneer
Saturday, 3 February 1883

The men in the mining district passed a law requiring that notice had to be printed in a newspaper to file a mining claim. S. W. Raymond quickly set up his printing press and started the *Animas Forks Pioneer*. A proud mine owner who lived just north of town in Mineral Point paid $500 for the first copy. After the first copy, the paper usually sold for a dollar, unless the paper was in short supply, and then it had been known to sell for $25. When mining started to slow down in 1896, the newspaper folded.

The Gold Prince Mine was the best producer, and the largest mining property. After the mine closed, the huge mill was moved to Eureka.

The Early Bird Mine and the Columbus Mine were good producers also. When production eventually declined, it slipped below the level needed to support the town. The Silver Coin Mine held on the longest before closing. The town emptied out in 1920.

In 1942 the train tracks were ripped up. Many foundations are scattered across this once beautiful town.

EUREKA
San Juan County

Directions: Eureka is nine miles northeast of Silverton on State Highway 110.

In the early 1860's a lucky prospector yelled "Eureka!" (Greek for "I have found it!") when he made a gold strike in this area. When the party of prospectors built some cabins, they named the area Eureka Gulch and the little campsite Eureka. At the time, Eureka was not as prosperous as they had expected, but by 1870 a few businesses and at least a dozen houses were built. A few area mines were worked: the Pride of San Juan, Cuba, Peters Placer, the McKinnie, the Yellow Jacket, and the Boomerang. The area newspaper was the *San Juan Expositor*. In 1873 the Sunnyside mine was found, and the next year the Sunnyside Extension. By 1882 *The Colorado Business Directory* reported a population of 150.

The Sunnyside mine went through several owners before Judge John Terry gained control of it. Some sources report that he got the mine for a few rounds of whiskey. A massive 500-ton mill was built, stair-stepping 18 stories up the mountainside. Five hundred men worked constantly, pulling the ore out of the mine, putting it in the tramway buckets, transporting it to the mill, and treating it.

In the summer of 1896 a rich vein of gold-bearing rhodonite was discovered, a rarity that had never been found anywhere else. This vein produced millions of dollars worth of the rare gold-laced rhodonite.

The same year, Otto Mears' Silverton Northern Railroad finally reached Eureka. At this time, Eureka and the Sunnyside Mining Properties were synonymous. Avalanches and rockslides constantly threatened the steep, treeless mountainsides bordering the gulches around Eureka. Otto Mears tried to erect a 500-foot tall, 10-foot thick wall to block the avalanches and keep the railroad open. The wall was wiped out that same winter.

In 1906 an avalanche carried away the entire bunkhouse for the Silverwing mine. This time, only one man died. With never any warning of pending danger, many people died this way. Despite its isolation, the town had electricity. In the summer, residents enjoyed long wagon rides, social dances, and picnics.

In 1910, the owner of the Sunnyside died, leaving the properties to his two sons and adopted daughter. The Terry children sold the properties in 1917. This was good timing because production had slowed. The post office for Eureka closed down in 1939, and the railroad was taken out later that year. The mine closed briefly during World War II.

In 1948 the Sunnyside faced bankruptcy. Its debts had accumulated to $3,600,000. The property was sold at auction for $225,000, and the massive Sunnyside Mill was dismantled and sold for scrap metal.

The town of Eureka is completely empty, with only ruins marking the town. People have tried to come back and work the mine, but these efforts have been unsuccessful.

GLADSTONE
San Juan County

Directions: Leaving Silverton, take State Highway 110 northwest for eight miles. At the end, State Highway 110 splits into four separate dirt roads. Take County Road 51, which will take you directly into the site of Gladstone after about a half-mile. This road is passable for a passenger car.

In the late 1870's a small mining camp called Poughkeepsie in Poughkeepsie Gulch was the first mining camp in the area. In 1879, a wagon road was built from Silverton up Cement Creek, leading to Poughkeepsie, three miles north of Gladstone, a new little settlement named for Britain's Prime Minister. At the time, Gladstone was building a chlorination works plant.

In 1882 the Sampson Mine was discovered. This was a good mine, and it employed many area miners.

One day in 1897, Olaf Nelson was working in the Sampson Mine, when he discovered a rich vein of gold. He followed the gold vein off the property and quickly staked his claim. Nelson did not have the money to develop his claim, so he worked it during his time off. He named his claim the Gold King. He dug a 50-foot shaft hole, and took out enough gold to make a decent living for himself until his death in 1890. Nelson left behind a nearly-destitute widow who in 1894 sold the Gold King to Henry Soule and Cyrus W. Davis for the meager sum of $15,000. Soule and Davis hired Willis Kinney as their mine manager. Kinney had been the one to suggest its purchase and had been waiting anxiously to develop the mine. In less than a year, a concentrator works was built, all of the claims in the area were bought, and tunnels and shafts were built and extended. Everything was incorporated as the Gold King Properties. The Gold King was a huge success, shipping hundreds of tons of ore a day. Hundreds of workers were employed, and rows of neat little white company houses were built. Boarding houses, saloons, and restaurants sprang up. The population mushroomed from 200 to 2,000.

In August 1899, part-owner and president of the Gold King Mine, Cyrus W. Davis, visited the mine, traveling half-way across the country from his hometown of Waterton, Maine, with a group of engineers. They built the Silverton, Gladstone Northerly Railroad. It was a big event for the town, a milestone.

In 1904 a fire swept through the mine buildings, trapping four miners in the tunnels. Rescue teams were sent in, but in the end rescue attempts failed and six men died.

In 1910, the stockholders and the heirs started to have problems, and the property was now thrown into litigation causing the mine to close down for eight years. This closure devastated the town, and many of the people left. In 1918 the mine was reopened with Kinney as manager, but this was only briefly successful.

In 1915 the railroad tracks were ripped up. The Gold King had produced between $8 and $9 million, but now production had slowed. The mine closed down in the early 1920's, and all the workers left again. The mine buildings and all of the metal buildings were torn down and salvaged for scrap metal during Word War II. Only piles of scrap lumber remain. The town of Gladstone is gone. A little bit of mining had been tried on and off, but this has been unsuccessful.

In the 1970's the Great American Tunnel Project was started, but it, too, failed.

MINERAL POINT (MINERAL CITY)
San Juan County

Directions: Mineral Point is 20 miles northeast of Silverton. Take State Highway 110 northeast and drive through Eureka toward Animas Forks. When you reach Animas Forks, stay on the trail on the right side of the river that goes over the town. Keep going. You will pass an intersection that goes to Cinnamon Pass; go straight past this. You will be heading to Engineer Pass. The trail will fork twice. Stay left both times. Soon you will pass a ridge leading to a high alpine meadow, swampy in places, lined with stunted pine trees. This is Mineral Point. You will need a good four-wheel drive vehicle, preferably a jeep.

While there are other routes, they are more difficult and are not recommended. One route goes from Ouray to Mineral Point, along the top of the canyon wall. Another route from Lake City crosses Engineer Pass. Take the dirt road on the right when coming down the top part of the pass. This road will take you by some mining remains before going across the meadow to Mineral Point.

As early as 1873, prospectors were in the San Juan area, near the headwaters of the Animas River. A prospector named Capt. Abe Burrows had made a good strike that year, calling it the Burrows Mine. Another prospector named Charles H. McIntyre had been busy locating lodes and building cabins. Both men tried to start a mining camp.

At an elevation of 11,474 feet, Mineral Point was a very isolated and cold town. It was named Mineral Point after a sixty-foot quartz "point," part of a huge quartz vein that ran on for miles with many small veins jutting off. Men started to come by the dozens to seek their fortune. Entrepreneurs in town created stories and drew pictures imagining the town's future transportation: steamships going up and down the Animas River, and trolley cars going to and from Mineral Point and Animas Forks.

The best mines were discovered in the first few years: the Ben Butler, the Bill Young, the Burrows, the Dacotah, the Mastodon, the Old Lout, the Red Cloud, the Vermillion, and the Yankton. Above the town was the very productive San Juan Mine and Mill.

The Bill Young was the best producing mine in the area.

The Mastodon Mine was in the quartz mineral point, for which the town was named, and it went through the vein for a few miles.

The mining company working on the Old Lout Mine was ready to give up. They had spent lots of money and had dug a 300-foot deep shaft without finding any good ore. One of the miners thought they should try one more explosion for curiosity. When the smoke cleared, they discovered a huge silver strike. That month, the mine produced $86,000 in silver ore. Unfortunately, ore had to be hauled out on very rough roads.

In the late 1870`s the population peaked at 1,000. The Mayor of Mineral Point, Ed Tonkyn, appointed himself postmaster, deacon of churches, operator of the Forest House Hotel, and street supervisor.

By the early 1880`s the mines grew less productive. *The Colorado State Business Directory* for 1882 listed the population as 1500. the town never recovered from the silver crash of 1893 and died a slow, lingering death. Almost nothing is left of Mineral Point. If you look carefully, you can find foundations or the ruins of the San Juan Mill above the town.

The Colorado State Business Directory for 1896 listed the following on Mineral Point:

MINERAL POINT
Mining town in San Juan county. Ouray, 8 miles, the nearest railway point. Daily mails. Population 200. Altitude 11,830.

Beatty, P B, mining contrctr
Bell, J H, loans
Calhoun, J H mining expert
Mineral Point Mining and Reduction Co
Orth, G S, chlorination wks
Richter, F, genl mdse & postmaster
Trickey, M M, assayer.

The *Animas Forks Pioneer* printed the following on Mineral Point on 22 July 1882:

$AN JUAN$
MINES JUST ROLLING OUT THE ALMIGHTY DOLLARS.
REPORT FROM THE NEW STRIKE AT MINERAL POINT.
And Hunt`s Placer Promises Bonanza of the best kind.

On July 1st, Messrs., Hunt, Kimber, and Harris located what is now known as the "Hunt Placer," a claim of sixty acres, located on the Animas river about a quarter of a mile below town. They constructed a dam upon the side of Treasure mountain about 1,500 feet from the river. The boom or reservoir is forty feet wide and fifteen feet deep, and with this they have prospected the placer by booming. The ground has shown good colors from the grass roots, and dirt was panned at a depth of eight feet showing splendidly, on Wednesday last, Messrs. Hoyt, Clark, Murray and the PIONEER man, escorted by Mr. Hunt, visited the placer workings. Just below the dam a fissure vein has been exposed by the washing waters, which has been staked by the Hunt party, and named Mecca. At the workings several pans of dirt were washed which showed 150 to 200 colors to the pan, the gold being small grains, with several particles as large as bird shot. Mr. Edwards exhibited to us the other day the result of a single panning, in which were a dozen colors as large as shot, and two or three regular nuggets. They are now working for bed rock and as depth is gained the ground looks better. As the now shows, it is confidently asserted that it will pay $10 per day to the man. They have surveyed for a flume, which will carry water from the river with sufficient fall to work the entire placer ground, they are working it right along, greatly satisfied and highly pleased with the prospect. The placer is admirably situated for development, and there is no reasonable doubt but that it will pay the owners handsomely.

The Animas Forks Pioneer, Saturday, 12 August 1882 printed the following on the mines in Mineral Point:

MINERAL POINT—On the Maude S., assays from a 10-foot cut gave 227.3 ounces of silver and 2.8 ounces of gold per ton..... There has just been completed $700 worth of work on the John lode, a shaft 5 by 7, 80 feet deep, which shows mineral the full size of the shaft..... The Napoleon No. 3 is showing finely- a 14 foot vein, highly mineralized with an 18 inch streak of extra good mineral..... On the Bill Young mine the development contemplated will consist of sinking the 80 foot shaft and running a level all the season. Hoisting works are on the road.

ALTA
San Miguel County

Directions: Go west from Telluride on State Highway 145 for about four miles. Before Keystone, turn left, going south on State Highway 145 for five miles. Then take Forest Route 625 left going east and follow this dirt road for a few miles. The road will fork after a couple of miles. Take the left fork. This will take you to Alta, just below Alta Lakes.

In the late 1870's, the Alta and the Gold King mines were discovered. The Alta Mine was a good mine, but it was the Gold King Mine that was the main producer in the area. At the same time, a little town formed on a high sloping alpine meadow below Alta Lakes. Named Alta, the little town was about a mile away from the mines. Almost all of the men worked at the Gold King Mine, which produced gold, silver, lead, and even copper.

The population peaked at 300. Although the town never had a post office, there were many cabins, buildings, a boardinghouse, a large tramway system, a school, and mine buildings.

The Black Hawk Tunnel entrance is in the town of Alta. The tunnel is 9,000 feet long and leads to the Alta and St. Louis veins. Alta was a very busy, prosperous little town. The mines produced at least $20 million. The mine and the town properties went through many owners: Four Metals Company of Milwaukee, Belmont and Tonopah Mining Company, Belmont and Wagner, Mr. H. F. Klock, and Mr. John Wagner, who on his deathbed married his housekeeper, leaving her his fortune.

Alta had a few different mills. In 1945 under the last mill that was built, an accidental explosion ignited a raging fire in the depths of the mine. The only way the superintendent could stop the fire was to blast the entrance of the tunnel closed, shutting off the oxygen. The seven men in the tunnel below tragically died. One was the son of the superintendent. The grieving superintendent never recovered from his son's death.

The mine was closed, and the town abandoned.

AMES
San Miguel County

Directions: Go west from Telluride on State Highway 145 for about four miles. Turn left, going south, before Keystone on State Highway 145 for about six miles. Then turn right going west on Forest Route 625 for half a mile. Ames will be on your left.

In 1882, Ames was a small town that formed around the Gold King's smelter. Ames was on the San Miguel River. The river was on the bottom of a canyon, and transportation was a problem for the Gold King. Money was also becoming a problem for the mine. At an altitude of 12,000 feet, the mine did not have any electricity. Miners had to use expensive fuels to operate their mining machinery. In 1882 *The Colorado Business Directory* listed the population at 50

L.L. Nunn was the attorney hired by the Gold King to help the financially strapped owners pay their creditors. L.L. Nunn analyzed the situation, and instead of liquidating and selling everything, he devised a way of harnessing the rushing water of the San Miguel River in the town of Ames. Nunn decided that they should build a power plant in Ames, and then transmit the electricity over high-tension lines strung over Imogene Pass at an elevation of 13,000-feet, right to the Gold King Mine. The mining engineers laughed at him, but Nunn went ahead with his plan. It worked, and the mine owners no longer had to use expensive fuels to run their machinery. Now they were able to make a good profit, pay off their creditors, save their mine, and save the town of Ames. L.L. Nunn had invented the first commercial transmission of high-pressure electricity, an application soon used worldwide. L.L. Nunn was a genius. He reportedly made his first fortune when he installed a big bathtub in his log cabin and charged prospectors, on their way to Telluride for a weekend of partying, 50 cents a bath.

Eventually, transportation problems and financial woes forced the smelter and the Gold King to close, and everyone moved away from the little town of Ames.

PANDORA   (Newport)
San Miguel County

Directions: From Telluride, take the main road out of town going east for five miles. At the bottom of Bridal Veil Falls, about half a mile away to the south, is the site of Pandora. This is at the base of Imogene Pass Road, which is closed, so you will have to hike into the town site.

The Pandora lode was discovered in 1875. By the next year, prospectors from Newport, Kentucky, started a small town, calling it Newport. It was below Bridal Veil Falls to the south, then called Waterfall Gulch. The scenery is spectacular. Soon the Smuggler-Union Mill was erected next to the town. It treated the ore from the Smuggler-Union Mine above the town. A huge aerial tramway system carried the ore down, and also transported the workers back and forth.

The name of the town was changed from Newport to Pandora after the Pandora Mine in 1881.

There were only 25 citizens the following year according to the 1882 *Colorado Business Directory*:

NEWPORT
Mining camp in Ouray County. Called Pandora.

Davis, E. L., grocery.
Pandora House, Mrs. Davis, propr.
Pandora & Oriental M. Co., (projected) 40 stamp mill.

In 1885 *Crofutt's Grip-Sack Guide of Colorado* listed the population at 60. All of the residents were miners, and the mail was brought in from San Miguel.

The little town grew and prospered, despite the threat of deadly avalanches every winter and spring.

The mines stopped producing in the mid 1900's. Today, the Idarado Mining Company owns the town and mining properties, and is working on a land reclamation project to restore the land.

PLACERVILLE
San Miguel County

Directions: Leaving Telluride, on the west side of town, take State Highway 145 west for 19 miles. Placerville will be on your right. Placerville is 38 miles from Ouray. The original town site of Placerville is at the intersection of State Highway 145 and State Highway 62.

In March 1876, Colonel S.H. Baker led a party of prospectors that discovered gold in the sands of the San Miguel River. By summer tents had sprouted like weeds, and men lined up the riverbanks panning for gold.

By 1877 the town was platted out and named Placerville, after all the placer gold that was being panned out of the river. Placerville had a few cabins and tents were scattered everywhere. An enterprising young man named Smith built a store and saloon about a mile south east of Placerville. When the lots and parcels surrounding his store and saloon went up for sale, everyone picked up and moved to Smith's businesses.

Placerville remained a viable, small town for at least a decade.

In the 1882 *Colorado Business Directory*, Placerville was listed as a small mining camp:

PLACERVILLE
Mining camp in Ouray County, 16 miles from San Miguel and 38 to Ouray.

William Frank, postmaster and groceries.
Joseph, Ed., hotel and feed stable.

In 1885 *Crofutt's Grip-Sack Guide of Colorado* listed the population at 125, and the main occupation in town was placer mining.

The Denver & Rio Grande Southern Railroad came through town in 1890.

The gold in the area started to play out, leaving ranching as the main occupation in town. The town was split: half of the town raised sheep, and the other half raised cattle. The cattlemen hated the sheep raisers because the sheep devoured the prairie grasses to their roots, leaving the ground bare. One day the conflict erupted into an all-day gun battle, and three sheepherders were murdered. The townspeople were shocked at the consequences of this violence among neighbors and were able to work out their differences.

The Colorado State Business Directory for 1896 listed the following for Placerville:

PLACERVILLE

Mining camp in San Miguel county, ou [on] the Rio Grande Southern Ry., 26 miles south of Ridgeway. Mail Six times a week. Population 200. Altitude 7,200.

Bertie, Robert, saloon
Evans, S J, gen mdse and htl
Evans, Walter, postmaster and notary public
Porter Cattle Co.

In 1919 a fire destroyed most of the business buildings. A few buildings were rebuilt, and a cattle company moved in and stayed.

The once bustling little boomtown of Placerville is now ranchland.

TOMBOY (Savage Basin Camp)
San Miguel County

Directions: From the eastern edge of Telluride, Tomboy is five miles down Forest Route 869, at the end of the road.

In 1880, the Tomboy Mine was discovered between Marshal Basin and Savage Basin, almost 3,000 feet above Telluride. The little settlement that grew up around the mine had been called Savage Basin Camp. The camp was very high up and very isolated. The mine was not worked much at first because of its location.

The Tomboy Mine was a gold mine, and after the silver crash of 1893, the mine owners looked at the Tomboy with more respect. The Tomboy Gold Mining Company started to produce gold in surprising quantities. The town's name was then changed to Tomboy, and the population soon peaked at 900. There were general stores, grocery stores, a school, stables, houses, cabins, and a daily stagecoach.

Several mines grew up near the Tomboy. The Columbia Mine was located half a mile from Tomboy. A short time after that in 1894, the Japan Mine was started half a mile away. The Smuggler Mine was one mile away from Tomboy.

Near Tomboy and its area mines was a camp called the Jungle. The camp was a notorious red light district, with saloons, gambling halls, brothels, and cabins. The owners of the area mining companies tried everything to get rid of the Jungle, spending thousands of dollars trying, but nothing worked.

In 1894 Tomboy sold for $100,000, and only three years later it was resold for $2,000,000 to the Rothschilds of London.

The miners that worked in the Tomboy Mine were members of the Western Federation of Miners. On 31 October 1903, 100 of the miners at the Tomboy went on strike. The mine owners hired nonunion men to work in the mill. Neither side backed down. In November, the mine owners wrote Governor James H. Peabody asking for troops to break the strikers. The mine had to be shut down. It was not until January 1904 that help arrived, in the form of the Colorado state militia led by Major Hill. The militia took the miners out of town to Ridgeway, and told them not to return. The men did not come back, and the mine was reopened. Over the next 23 years, the mine's production fell every year, until the owners could not justify keeping it open. It finally closed in 1927.

Everyone in town moved away, and only ruins remain. As of 2000, the Idarado Mining Company owned the Tomboy town site, the surrounding 7,200 acres, and the mine.

ARGENTINE (Decatur, Rathbone)
Summit County

Directions: Argentine is 16 miles east of Dillon. From the eastern edge of Dillon take U.S. Highway 6, heading east and passing Keystone along the way, for six miles. At that point take Primary Forest Route 5 to the right for five more miles until you reach Forest Route 250 on the left side of the road. Take this for four miles to the original site of Decatur, which will be on the right of the road. The site of Argentine is one mile after Decatur.

In 1859, Stephen Decatur Bross abandoned his wife and two small children in Poughkeepsie, New York, and moved west to Georgetown, Colorado. One day he just walked out the door and never looked back. He dropped his surname of Bross and became Stephen Decatur.

In 1868, Stephen Decatur, after some silver strikes, established a town, naming it Decatur. Back east in Illinois, Governor Bross had heard that a man in the mining town of Decatur resembled his brother who had vanished in 1859. Governor Bross made the trip to Colorado, but when the two met, Decatur denied that he knew Governor Bross. Governor Bross returned to Illinois, grieving for the brother he had lost. Stephen Decatur returned to his usual business, always keeping busy.

In his early days fighting Indians, Decatur was the main builder of Argentine Pass, and he helped establish many little towns. In 1876 he represented Colorado in the Philadelphia Centennial Exposition, lost all his money, was crippled, and moved to the town of Rosita, where he lived on the charity from others and died penniless. Stephen Decatur is buried in an unmarked grave in Rosita.

Mining slowed for a while because the early silver strikes were low grade. This changed in the 1880's when the Pennsylvania Mine was discovered in 1879 by J.M. Hall and produced high-grade silver ore. The population of the little town soon reached 100.

The Colorado Business Directory published in 1882 listed the following information:

DECATUR
Mining camp in Summit County, 29 miles west of Georgetown. Daily mails. To Breckenridge, 25; to Kokomo, 20.

Decatur House, W. L. Sawtell, proprietor.
Delaware Mining Co.
Eyre & Lathrop, assayers.
Good Hope Consol. M.& S.
Heck, Casper, restaurant.
Kelley, J. P., mineral surveyor.
Lathrop, E. S., assayer.
Orphan Boy Mining Co.
Shadbolt, Geo., gen'l mdse.
Snowslide House, Mrs. A. J. Brickner.
Trull, J., blacksmith
Widitch, & Co., hotel and saloon.

By the end of the 1880's, most of the residents had left and the post office was closed.

When people returned to the town, they changed its name to Rathbone. The area mines picked up in production. The Pennsylvania produced around $3 million in ore. Others were the Delaware, the Peruvian, the Queen of the West, and the Revenue Tariff. The ores assayed at $20 to $2,000 a ton. Just northeast of the town was the Peruvian Mine, nicknamed the Peru Mine.

Peru, as it was called, was never officially a town, but a little settlement that grew up with a few cabins, a boardinghouse, a couple of buildings, and a few mining claims.

Sometime in the 1880's, two men from out of town bought a claim on the side of a mountain and hired a few miners to work for them. One of these miners was Gassy Thompson. The claim owners told the men that they needed a tunnel dug out of their claim a hundred feet deep into the mountain. Winter was coming, and the claim owners said that they would come back in the early spring when the job was done and pay the men. Until then the men had a warm cabin, supplies, and food to wait out the winter. Gassy Thompson and his assistant simply relaxed in the warm, snug cabin and waited for a deep winter snow pack to accumulate on the mountainside. The men dug a tunnel a few hundred feet through the snow. The two men rubbed the inside of the tunnel with dirt and put up a wooden beam here and there. In early spring, the claim owners returned to the area and inspected their mine. They were very pleased with the work and paid

Gassy Thompson and his assistant well. The miners took their pay and left immediately. The two naïve mine owners were very happy and proud of their investment. . . until April when their mine melted.

In the early 1890's the town of Rathbone died, and the post office was closed.

Not long after that, people moved back again. This time, they moved the town site to the other side of Peru Creek, and about a mile up the road and to the left. Argentine, named because it was at the foot of Argentine Pass, had a post office, a few stores, and a cluster of cabins. The closest railroad station was Keystone. Life in Argentine was good until the silver crash of 1893 forced most people to leave.

A few years later, people started to move back.

In the early spring of 1898, a massive avalanche thundered down the mountain and swept the whole town away, never to be rebuilt. It is very difficult to find remains of the town.

CARBONATEVILLE
Summit County

Directions: From Frisco take State Highway 91, to the site of Carbonateville, 12 miles south and on the left side of the road. Carbonateville is one mile southeast of the site of Robinson and two miles south from the site of Kokomo. Carbonateville is 18 miles north of Leadville.

In December 1878, the town of Carbonateville was laid out, following the discovery of rich silver. This was the first town in the Ten Mile District. Miners, prospectors, gamblers, and others filled the area. Soon, sturdy little log cabins replaced the makeshift tent city.

For the brief period from 1860 through 1862, early prospectors took out over $3 million in placer gold before the gold played out, and the prospectors moved on.

Sixteen years later, the area was rediscovered, but this time it was silver that was discovered. Neighboring Leadville was booming, and prospectors expected to find silver throughout the area, focused on finding it, and were successful.

By 1879, with the new silver hard rock mining, Carbonateville was a silver boomtown with over 60 businesses. The hotels in town included The American House, The Carbonateville House, The Joplin House, and The Oakland House. The local newspaper that everyone read was the *Ten Mile News*. A Leadville stagecoach provided transportation. Seven saloons kept the town rowdy. In 1880, other camps popped up, like mushrooms after a spring rain, within a four-mile radius. To the north were Kokomo, Recene, and Robinson.

After the railroad came to the town of Robinson, turning it into a major mining and supply center. When this happened, most of Carbonateville's residents moved to Robinson. By 1881, Carbonateville was deserted.

Today, the town of Carbonateville is blanketed by tailings piles left by the Climax Molybdenum Company.

CHIHUAHUA
Summit County

Directions: From the eastern edge of Dillon, take U.S. Highway 6 east, passing Keystone, for six miles. At that point take Primary Forest Route 5 right, for five more miles until you reach Forest Route 250, on the left side of the road. Take this for two miles. Do not take the dirt road that is just less than two miles; this is the back way to Arapahoe Basin Ski area. About a third of a mile farther is the dirt road on the left going up Chihuahua Gulch. Take this road. The site of Chihuahua will be spread out over the next couple of miles.

The mining town of Chihuahua, on Middle Snake River, in Chihuahua Gulch, was officially incorporated as a town in 1880. Of the high mountain peaks surrounding the town, the highest is Gray's Peak, just northeast of the town at an elevation of 14,270 feet. Many of the residents of Chihuahua worked at the Pennsylvania Mine, four miles down the road, and at the Peruvian Mine, by Argentine Pass. The ores being mined were mostly silver, with some gold, copper, and lead, paying out at $60 to $7,000 a ton.

The population reportedly peaked at 200. In 1885, *Crofutt's Grip-Sack Guide of Colorado* reported that 100 people lived in Chihuahua.

The business district was along the town's main street. The Snively and the Chihuahua were the town hotels. There were also a few stores, a sawmill, and a school just east of town was shared with the town of Decatur, a mile and a half away. Chihuahua had its own reduction works, but there was no church or doctor. The townspeople said there was no need because this was healthy town without crime.

The townspeople were right, there was no crime in early Chihuahua. Then, two prospectors were robbed and murdered in the woods. The townspeople soon discovered the bodies. They assembled a posse, chased down the killers, and captured the three of them. The posse hung the murderers in the woods. This was boomtown justice. Then they say that the townspeople dug two graves on the outskirts of town: one for the prospectors, and one for the murderers.

In the spring of 1882, *The Summit County Times* reported:

At Chihuahua the old timers are coming in on every stage and business is increasing with the merchants. Crilly & Co. are laying in a large invoice of staple groceries and miners supplies. The roads are yet in bad condition, not owing to quantity of snow, but the sidling condition of the road bed. The toll company are at work with a force of men between Chihuahua and Montezuma, and by the first of the month wagons will be running without hindrance. Frost & Co. who have very fine prospects up Chihuahua gulch, placed the properties to a Boston outfit this winter this winter, and expect to put about twenty men to work in a short time. A tunnel is being run to develop the Brittle Silver on Silver Mountain. A force of ten men are working on it with flattering indications for the near opening up of a fine ore body. The Brittle Silver is a promising property at the surface. It shows a well defined vein twenty-six inches between walls, with a five inch pay streak running 3200 ounces per ton. The properties of this section are invariably high grade fissure veins in granite.

The Summit County Times
Saturday, 29 April 1882

In 1889 a forest fire swept through Chihuahua, destroying the town. The town was never rebuilt. The townspeople left and never came back. Nothing was left but ashes and burnt timbers.

Looking at the area today, you could not imagine that anyone had ever lived here.

CONGER (Conger Camp)
Summit County

Directions: From Breckenridge go south on State Highway 9 for about a mile until you reach Boreas Pass Road, which is also called County Road 10. This is an old stagecoach road that went from Breckenridge to Como. Take this road. No obvious trace of the town is left, so it is hard to find. Reliable sources differ on the location of Conger. *Croffut's Grip-Sack Guide of Colorado* published in 1885, reports that Conger is three miles south of Breckenridge on Boreas Pass Road. A local Summit County historian states that Conger is five miles south of Breckenridge on this road. Conger is at the base of Indiana Gulch in a half-moon shaped valley.

In the late 1870's Colonel Sam Conger, who had found the Caribou Mine in the 1860's, located the seemingly rich Dianthe Mine. This rich shallow fissure vein was spread like a blanket of silver over the area. Assays on the ore ranged from $1,000 to $2,000 a ton. By 1880, a year after the discovery of the Dianthe Mine, about 40 houses had been built, along with a general store, a post office, a sawmill, and a few other main street businesses. Many other good strikes were made, both placer and lode. The best were the Case, the Franklin, the Highline, and the Newark City. In the beginning, the men were unable to find investors to develop their property.

It is said that the town had had plenty of prospectors to find the good areas, but not enough miners to mine them once they were found. Maybe that is why it took the prospectors so long to learn that these rich fissure veins would play out soon, once they were mined. By the time the prospectors found investors to develop their property, their property was played out. The mines produced brittle and ruby silver, silver glance, and copper. As the mining slid, the men turned to lumbering. The town of Conger was surrounded by heavy timber in Indiana Gulch. Deer, elk, mountain sheep, and trout were plentiful for game. Occasional sightings of mountain lions and grizzly bears kept the men on the alert.

The Spottswood and McClellan Stage Line linked the town to other areas and brought mail into town. In 1882, the railroad passed about a mile from town.

The lumber business and the railroad were not enough to save the town, and the people left. There is nothing left.

DELAWARE FLATS (Delaware City, Preston, Braddock)
Summit County

Directions: Coming from Dillon, take State Highway for 10 ½ miles, until you reach Tiger Road on your left. Take Tiger Road for a little less than a mile until you reach Forest Route 300, which is also called Gold Run Gulch. Delaware Flats, the name given the community founded on swampy flats of the Swan River, was located on the right.

A mining camp grew up near the junction of Swan River and Blue Creek. By 1860, the town of Delaware Flats was established. This was one of the earlier towns in the area, with placer mining the town's main support. Delaware Flats had at least 30 houses by 1861. That November a post office was established with George Ander as postmaster. Below the base of Gold Run Gulch was the little town of Buffalo Flats.

Delaware Flats died out twice, and it went by a couple other names. In the 1870's the town died out and came back as Preston. Then in the 1880's the town died out again, but it was revived and called Braddock, after Dave Braddock, and his Broncho Stage. The locals continued calling it Delaware Flats, and that was the name that stuck. The change of names made things confusing for everyone since there was another town named Preston at the south end of Gold Run Gulch, and another town named Braddock off State Highway 9 halfway between Dillon and Breckenridge *The Breckenridge Daily Journal Newspaper* printed the following vignette on Preston on Friday, 7 June 1881.

The town of Preston comprises a dozen or so miners cabins and mining companies' headquarters. The postmaster, John Shock, is well known as a pioneer in Summit county. He is a miner, prospector, newspaper correspondent, and mine owner. In his opinion he has some of the best mining prospects in Summit county and exhibits his faith in the future by doing the yearly assessments, keeping his residences and predicting the future glories of Gold Run. It is but fair to say of Mr. Shock, that he has done yeoman's service for this section of the western slope when it had few friends, he has served its best interests and deserves well for his efforts.

During the winter of 1880-1 a wagon road was projected from Breckenridge to Preston and is now passable for light teaming, the distance is less than three miles, and a small expenditure would open this section of town. The sum of $75 was contributed by the citizens of Breckenridge.

The Breckenridge Daily Journal
Friday, 7 June 1881

The Colorado State Directory for 1882 listed the following for Preston:

Preston
Mining camp in Summit County, 3 miles northeast of Breckenridge, and 2 ½ miles from Lincoln City. Tri-weekly mails.

By 1890, mining in the area slowed, the post office was closed, and when the residents moved away, Delaware Flats was abandoned. A few men did a little mining until the 1930's in the surrounding area, but it was not enough to save the town. There is nothing left of the town.

In recent years a few houses have been built in the area, and soon there will be more. Reclamation projects have restored some of the land in this area, cleaning up residue of cyanide used by the mines to recover gold.

DYERSVILLE
Summit County

Directions: From Breckenridge, go south on State Highway 9 for about a mile until you reach Boreas Pass Road, which is also called County Road 10. Take this road. You will see a Forest Service sign when you are halfway between Breckenridge and Boreas Pass. The sign directs you to Indiana Gulch off the main road. Take this, and at the head of the gulch is Dyersville.

Off Boreas Pass Road lie the ruins of the Seven Forty Mine, and a four-wheel drive road goes downhill from there to Dyersville.

Below Boreas Pass, about two miles northeast, is where the little town of Dyersville sprang up in the early 1880's. The town was at the end of Indiana Gulch, close to Indiana Creek. Dyersville was named after its founder, Father John Lewis Dyer, who had devoted his life to preaching his Methodist faith to anyone who would listen. Walking from town to town and preaching on street corners, he earned the name the Snow-Shoe Itinerant. The Methodist Church of Denver paid him a salary of $125 per year to preach and to bring the Methodist faith to the Breckenridge and Blue River area. Unfortunately, room and board alone cost almost $10 for just one week. Not letting anything stop him and not having enough money to eat, he accepted a job delivering the mail, treking from Buckskin Joe through California Gulch to Cashe Creek. This route enabled him to hold a regular church service at Cashe Creek. He made this forty-mile trip, carrying about twenty-five pounds, and in the winter climbing through snowdrifts sometimes twenty feet deep, for a salary of $18 a week. He stopped along the way and helped anyone in need. If anyone needed to talk, he was always there to listen. On one of Father Dyer's trips, he discovered some rich ore. Thinking that maybe this could support his religious mission, he filed a claim, calling it the Warrior's Mark Mine. Not having enough money to have his ore assayed, Father Dyer convinced a Breckenridge assayer to assay his ore in exchange for a partnership in the mine. News spread of Father Dyer's strike, and other people came seeking their fortunes, staking claims, and building cabins. Soon a general store, a restaurant, a saloon, a school, and more businesses were erected. Father Dyer built a few cabins himself, and then he moved his faithful loving wife to the camp. This was in 1881, and Father Dyer was just turning 70 years old. After about a year or so, he was feeling that he was getting a little too old for the mining business and sold his half of the mine for $2,000. He returned to preaching and ministering to his flock.

On 19 January 1890, *The Denver Republican* reviewed Father Dyer's autobiography, providing a glimpse into the life of this remarkable pioneer preacher.

FATHER DYER`S BOOK
The Aged Preacher Writes an Autobiography of Rare Interest

"The Snow-shoe Itinerant," an autobiography of Rev. John Dyer of the Colorado Conference, Methodist Episcopal Church, is by far the most unique and withal sensational book of its character, that has appeared for many a day, and cannot fail an immense sale in Colorado, where "Father Dyer" has been known since the memorable days of `61, as a pioneer preacher, pilgrim, and patriarch. The volume is just from the press, and a perusal of its 360 pages sends the blood tingling through the veins of the reader, awakening recollections of trail and campfire that no other recital could accomplish. Everybody in the West knows "Father Dyer," but not everybody has the faintest conception of the story of the Western life embraced in a third of a century labors, and exciting episodes of missionary work on the frontier and amidst the rugged crests of the Rocky mountains.

Father Dyer tells in his homely yet inimitable way how he walked from the Missouri river to Denver in 1861, and his quaint description of the struggling village on Cherry Creek with its strange types of Western life, is like a breath blown from the tips of the Colorado peaks. In Colorado and New Mexico "Father Dyer" laid broad and deep the foundation of the Methodist church of the West, traveling 10,000 miles on horseback in two years, climbing precipitous mountains, crossing terrible deserts, living with Indians, facing desperadoes, suffering hunger, cold, and thirst, enduring all for the sake of the master and the church he loves so well. The venerable author is the father of Judge Elias Dyer, who was assassinated in his court room at Granite, Lake county in July, 1875, and several pages are devoted to an account of the tragedy, giving several hitherto unpublished facts in relation to it. Father Dyer walked from Summit County to Denver in 1861 to attend the first ministerial associates ever held in Denver. Afterward he read a paper at the association, giving an account of his first year in Colorado. Governor Evans gave him $15 for the paper after it was delivered. But it is useless in this short sketch to attempt to give a faint impression of what the book contains, or what manner of man is this who in a simple and almost childlike way narrates incidents that would adorn the pages of Verne or Du Chaillu. And their intensity lies in the knowledge that they are true. Next to the biography of Peter Cortwright, says Professor S. W. Williams, "this is the most interesting in the annals of American Methodism." It is not Methodists alone, however, who will read its pages and not grow weary. The old frontiersmen and mountaineers and miners will find within its covers much to recall the scenes, which the pioneer preacher has so vividly depicted.

The Denver Republican
19 January 1890

The town of Dyersville prospered throughout the 1880's, but eventually the area mines played out, and the townspeople left.

Only ruins remain.

KOKOMO
Summit County

Directions: Kokomo is 296 miles from Denver and 18 miles northeast of Leadville. Leaving Frisco on State Highway 91, the site of Kokomo is ten miles south on the right side of the road. The site is covered by tailings piles left by the Climax Molybdenum Mining Company.

In 1878, during a spring snowstorm a party of Leadville prospectors, originally from Kokomo, Indiana, were traveling through the Ten Mile Creek area. In the bitter cold one of the men froze to death. He was nicknamed "Scotty" because he was Scottish. The group set up camp and arranged for one of the prospectors to bury him. The next day the men noticed that the man who was supposed to dig the grave was gone, but he had left a note behind. The men read the note in disbelief. "Dug four feet down, struck it rich, went to town to file claim." That discovery triggered the start of Kokomo.

Kokomo was officially incorporated as a town on 3 June 1879. The founders named it after their hometown of Kokomo, Indiana. The new Kokomo was located on the east side of Sheep Mountain at an altitude of 10,618 feet. Prospectors came seeking their fortune and many claims were filed.

Many other little towns were located near Kokomo. The town of Recene, founded by the Recene brothers, sprang up next to Kokomo to the south. The town of Robinson sprang up in 1880, two miles south of Kokomo. The town of Carbonateville was between Climax and Robinson. A few more little towns and mining camps were located within this four-mile radius. This was called the Ten Mile Mining District.

On Saturday, 18 June 1881, *The Breckenridge Daily Journal* reported a killing in Kokomo:

Killing at Kokomo
Special Dispatch to the Journal

KOKOMO, June 17.- Al Huggins, a well known desperado of Ten Mile, with Phil. Foote, ex marshal of Kokomo, who is not a bad character except when under the influence of liquor, have created a great furor in the pioneer camp. At noon today they came up from Recen[e] in a quarrelsome state of drunkedness and began firing their revolvers at random; shortly after encountering Mayor Donacaster, of Racen[e], on the street, Huggins deliberately fired his revolver at him, the ball just grazing his head. At this instant Policeman Brown stepped up and mildly insisted that Huggins be quiet as there was no necessity for a quarrel. Huggins, with drawn revolver bestowed the vilest epithets on the officer and fired, the ball striking Brown just below the collar bone and near the first rib. Brown is in a very dangerous condition, though he may recover. Huggins and Foote fled to Recen[e] where they collected arms and prepared for open resistance. Marshal Sutton immediately engaged a posse of men and going to Recen[e], the suburb, shot Huggins in the face with a load of shot, he then with Foote gave himself up and both were incarcerated in jail. This afternoon the organization of a vigilance committee became evident to the officers and they hurried the prisoners to Robinson where they took the evening train for Leadville. There is much excitement in camp, as Brown was an excellent man and well liked.

LATER.
KOKOMO, June 18.- Foote escaped from the officers last night while on their way to Leadville.

The Breckenridge Daily Journal
Saturday, 18 June 1881

The population of the area reportedly peaked at 10,000 in 1881. The area had continued to grow until the fall of 1881, when a fire raged through Kokomo, destroying the city. Almost everyone left. *The Breckenridge Daily Journal* reported the details of the fire, along with estimates of the losses. These loss estimates draw a picture of the thriving community and its residents.

KOKOMO DESTROYED
The Town Wiped out as With the Besom of Destruction
The Latest Particulars By Telegraph and Private Sources
Want of Water the Cause of This Wholesale Loss
A Large Number of Families left Shelterless
A Partial Statement of Individual Losses
The 13th of October a Day Long to be Remembered

Fire was discovered in the Summit House about seven o`clock and was caused, it is supposed by the accidental explosion of a kerosaene [kerosene] lamp. The fire followed so close on the explosion that Mr. Longwell, of the hotel, had not time to take the money out of the drawer, he lost everything including a new $4,000 outfit just purchased. The flames spread with great rapidity on both sides of the street. The Summit County bank was first attacted and went down like timber. About the same time the Morris` dry goods house fell prey to the fire fiend, which also went down like a flash, nothing being saved. Mr. Morris` family was removed to safety. The Western house was the next victim. Mrs. Strong saved a trunk or two, but Strongs horses were burned and everything about the premises. The flames next lapped back and devoured all that was supposed to be safe on the north side of the street and continued down the hill, burning out Huss & Smith`s saloon, Sheperd's law office, Staats' assay office and fruit store, some vacant buildings, Harder`s grocery, the post office, Stainsby`s drug store, Boyd & Co.`s shoe store, Stainsby, Fort and Colby`s hardware and general store. A large frame building was torn down, which arrested the flames on this side of the street. From the bank, the flames spread rapidly along Ten Mile avenue taking in Georges saloon, Wilkins' saloon, Hill & Butler`s law office, a fruit stand, Wesley`s livery stable. Mr. Wasley was fortunate in saving his stock, a dance house, a vacant building, Hyman`s Georgetown mint, total loss, several vacant buildings. The Summit County Times was a total loss, the Splendid power press being entirely destroyed, the buildings on the opposite side were also destroyed. From the Western house on the lower side of the street to Hines & Dowd's grocery store everything is burned out. Bengly`s tin store was saved with hard work. The losses are variously estimated from $250,000 to $400,000. A meeting was held at the M. E. church, Friday morning, when the spirit of the citizens was shown in a determination to rebuild and not allow this misfortune to interfere with the growth of the town. Telegrams were at once sent to Leadville and upon receipt of intelligence of the fire, General Agent George D. Williamson of the Rio Grande, referred the message to Superintendent Griffin, and a special train was at once ordered, which was immediately filled with firemen and citizens. The train was in the charge of Conductor Sam Black,

Engineer J. McCale and Fireman L. E. McCale, with Superintendent Griffin, George Williamson and Master Mechanic John Walker on board. Instructions were given to send the train through as fast as safety would permit. The Tabor hose, Humphry hose and Harrison Hook and Ladder companies were largely represented and did all they could to assist. The smelters and mining property escaped, and no one was injured. About eighty families were rendered homeless. Owing to the confusion it is impossible to give a complete list of losses, but we compile from various sources an approximate idea of individual losses:

Windsor hotel, $1,000
Union hotel, $1,500
Summit house, J. M. Longwell, proprietor, $5,000
S. L. Morris & Co., dry goods, $12,000
Several small buildings, aggregating $6,000
J. N. Harder, groceries, $4,000
Post-office, $2,000
Stainsby, Fort & Colby, $6,000.
Several offices and one saloon, $1,500
Sipple Bros., grocery store and liquors $8,000
On the east side, Dan Davies, butcher, $2,000.
Times office, $6,500.
J. A. McCune $1,500.
M. Hyman, saloon, $3,500.
E. Weinmerstadt, $1,500.
L. R Dean, $1,500.
Wm. Wesley, livery, $3,500.
Brookly & Alexander, $2,500
H. T. Wilkins, $3,500.
Geo. R. Fisher & Co., $4,000.
Hill & Butler, $1,000.
Geo. Bengly, $1,500.
R. K. Smythe, $6,000.
Western hotel, $5,000.
Staats, $2,000.
Sheperd, $1,000.
Judge Morrison, $1,500.
Calhoun building, $3,000.

Thos. Killduif, $4,000.
Dublond saloon, $1,500.
Winkleman, $5,000.
Odd Fellows and Masonic Hall, $500.
Jewelry store, $1,500.
R. Heitler, $25,000.
Chas. Wendler, $3,000.
Thos. Latter, $4,000.
D. F. Mitchell, $1,500.

At the earliest possible occasion the full particulars will be given, but as the line is crowded with private and commercial dispatches which prevents an extended telegraphic notice. The Journal is indebted to Messrs. V. F. Veis and G. W. Lindgreen for the graphic account of the fire and to J. M. Longwell for assistance in revision of the losses. Among the most exciting incidents of the fire was the explosion of powder, cartridges and explosives, which occurred in all the general stores. Fortunately no one was injured.

The Breckenridge Daily Journal
15 October 1881

Following the near total destruction of Kokomo, the neighboring community of Recene offered land to the homeless refugees:

The Kokoma-Recene Removal

The following is the proposition which the Recene town company has made to the citizens of Kokomo: "We hereby agree to donate lots in the town of Recene to all residents of Kokomo who were carrying on business and were burned out at the recent fire, in consideration that the parties to whom such lots are donated shall erect buildings to cost at least $300, said buildings to be begun within thirty days from the date of selection of lots and continue to completion without unnecessary delay. The town company will receive at their office in Recene applications for lots for the period of ten days only from this date, but they reserve the right to select lots for the various applicants, as in their judgement the demands of business require. Bonds for deeds shall be given on selection of lots, and said bonds shall not be transferable. Recene. October 15, 1881." It is understood that if the time of building (thirty-days) mentioned in the above proposition is found to short, any reasonable extension will be given. We hear that a lot intended to be donated for church purposes has already been selected, also lots for the Masonic fraternity and Odd Fellows. Some of the business men of Kokomo were so anxious to secure eligible lots in Recene, that they went to the town company and bought such as they desired. Among those who had filed their applications for lots early on Monday last, were The Times, the post office, Robert Gut, J. N. Harder. Charles Wendler, Thomas Latta, and various others. The lots in Recene are 25 feet by 100 feet in extent. The avenues are 60 feet wide and the cross streets 50 feet. Through each block runs an alley 15 feet wide. Pollock avenue will no doubt become the main business thoroughfare of the new town, for it is just half the way between the two proposed depots—those of the D. & R. G. and South Park. The D. & R. G. has already some buildings in Recene.

The Breckenridge Daily Journal
Wednesday, 26 October 1881

In 1882, *The Colorado Business Directory* listed the population of Kokomo at 200 and reported that the Denver & Rio Grande Railroad went through the town. The town was rebuilt and joined with Recene, but it was never the same.

By 1885, the population had grown to 1,000 according to *Croffut's Grip-Sack Guide of Colorado*. The community had two local newspapers: *The Summit County Times* and *The Times*. The Summit and the Western were the town hotels.

Kokomo was still recovering from the devastating fire, when it was again struck down by the silver crash of 1893. The town's population slid to about 200, and Kokomo held on as a company mining town.

The Colorado State Directory listed the following on Kokomo in 1896:

KOKOMO

Mining camp in Summit county on the D. & R. G., U. P. D. & G. Rys. Population 350. Altitude 10,500. Distance to Leadville 18 miles.

Arnold, P R, meat market
Bryant, Joe, saloon
Clifton House, Martin Benson , prop
Colcord, A B, city clerk and notary
Col Sellers Mine, G T McDonald mgr
Dowd, J W, genl mdse
Johnson, J E, physician
LaFountain, E, groceries & meat market
Mountain House, Pangburn & Cuthbert, props
Paine, Thos, barber
Plandett [?], T F, Uncle Tom group of mines
Quackenbush, G W, hotel
Shaw, blacksmith and transfr
Summit Mining and Smelting Co. E E Byron, pres
Thompson, G R, assayer
Thompson, W A, drugs and clothing
Tracy, W C, groceries
Whitney, S W, saloon
Wildy, A R, concentration works

In the early 1970's the Climax Mining Company tore down the remaining buildings, and covered the site with tailings piles.

LINCOLN (Paige City, Lincoln City)
Summit County

Directions: From the eastern edge of Breckenridge, take French Gulch Road for four miles, and you will be in Lincoln. Two sources report that the original site of Lincoln is a mile away from the later known site, but we do not know in what direction.

In the early 1860's, prospectors filled French Gulch, at least 200 men were placer mining along the river four miles east of Breckenridge. The coarse placer gold was pure and very plentiful. The prospectors were making a very good living for themselves and were quite satisfied.

Harry Farncomb had come to French Gulch in 1860, and had been placer mining here for 19 years. In 1879, Harry Farncomb went upstream to find the source of his coarse placer gold. Not finding any visible traces of gold, he started to dig shallow holes in the shale rock on a hill that was in upper French Gulch, five miles east of Breckenridge. This is how Farncomb found "the wire patch." He discovered large amounts of crystallized gold that was made up of masses of wires, and he was also found sheets of delicate leaf gold. Farncomb was very careful not to let anyone see what he was doing. He bought up the property that held his wire patch, but he was afraid to file a claim because that would have aroused suspicion. He spent much of his time digging holes and sluicing out the dirt. Farncomb soon discovered the tops to several gold veins in his hill. He kept his gold well hidden and kept accumulating it, until finally in 1880 he went to Denver and deposited 300 ounces of his beautiful unique wire and leaf gold. Word spread fast of the discovery, prospectors and even regular working men from Denver rushed to French Gulch to get rich, too, only to discover that Farncomb owned all of the property. The disappointed prospectors were angry, jealous, and accused him of cheating his fellow man by being so selfish. Farncomb held his ground. Some men from Denver tried to legally gain control of Farncomb's property. Farncomb owned the property with the gold wire patch, but he had never filed a claim. This was the beginning of the "Ten Year War." The battle in court cost many hundreds of thousands of dollars and bankrupted a Denver bank. Still, Farncomb kept raking in the gold, literally. The men who were using the law to grab Farncomb's Hill were, so far, unsuccessful. With their patience – and funding -- running out, they decided to take the property by force and went to French Gulch with guns drawn. The shootout lasted seven hours, many men were seriously injured, three men were killed, and at the end of the day nothing was resolved. During the last two years, he had mined 7,000 ounces of gold from his hill. Eventually Francomb sold his property to a neutral party and retired a rich man.

The Breckenridge newspaper, *The Daily Journal*, provides the clearest portrait of Lincoln during the early 1880's.

LINCOLN
Our lively suburban neighbor - placer and lead mines
Roads and prospects for business—a promising outlook

Last week, a representative of the *Journal* spent a couple of days at Lincoln and in its vicinity visiting mines, conferring with citizens and noting the signs of healthy improvements to be seen in all directions. Lincoln is located and situated that is in every respect, financial, mercantile and socially, tributary to the county seat. Almost every business man and mining operator in Breckenridge has some interest in the many mining properties in the vicinity of Lincoln, while all the supplies are purchased here and financial transactions conducted through the bank of Breckenridge. Thus it is, that writing of Lincoln and its enterprising inhabitants we are commenting on our own people.

The distance from town is generally over rated, while it is called four miles, the distance is really but a trifle over three and would be a delightful ride at all seasons of the year, were it not for the horrible condition of the road, which would be a disgrace to the uttermost, frontier town. This route leading from French pass and Georgia pass along French gulch to Breckenridge is one of the principal approaches by highway to the county seat and is traversed daily by scores of people yet, is in such a condition that teams are liable to be stuck coming down hill.

It is hoped and expected that the county commissioners will occupy a small portion of their valuable time in rectifying this state of affairs, as there is enough voters polled, along the route to keep the road in excellent condition.

Lincoln is an old town having been the scene of the wildest excitement ever witnessed in the Rocky Mountains over the discovery of gold in wire patch and Jeff. Davis' patch, both in the limits of the town.

The foundations of extensive buildings are still to be traced in the vicinity and hydraulic ditches are laid out on a scale to astonish a modern placer operator. Even now, drifting in the banks of French gulch in winter and ordinary placer mining and hydraulic work is profitable when properly conducted. During the winter two men have averaged $10.00 per day working together, putting in perhaps eight hours daily and doing their own cooking. Last week Perkins and son took out $17.50 as the result of a day and a half.

. . . It is reported that others have done equally well. There are thirteen men now engaged in drifting, which is the average number that have been at work during the winter.

Messrs. Smith and Cooper, have a lease on considerable ground and will work a number of men soon as the weather will permit. George and Harry Farncomb will operate their extensive hydraulics, placer claims, and lead mines, with as large a force as can be economically employed. Church and Cook are drifting on good ground, on which they have a lease to May 1st. They expect to obtain leases on considerable property and work a number of men during the season. Calvin & Geo. Clark are expected to work their extensive claims during the season together with many others now absent for the winter, while the snow fall this winter has not been great as in past years, advantage will be taken of all the water that flows through the gulch.

The Daily Journal
21 March 1882

The Colorado Business Directory published in 1882 reports that Lincoln City had a population of 105. A few years later, in 1885, *Crofutt's Grip-Sack Guide of Colorado* listed the population at 250. Other sources say that the population peaked at over 1,500 in the 1880's. The two hotels serving the town were The Wheeler and The Perkins. There was one general store, a few hydraulic mining companies, a stamp mill, and one steam saw mill. Some of the area mines were: the American Union, the Bismark, the Cincinnati, the Elephant, the Governor King, and the Queen of the Forest.

As the gold production decreased, people drifted away.

Today, Lincoln City is gone, and only a few people live in the area.

MASONTOWN
Summit County

Directions: Mason Town is remote and difficult to find. It is between a half mile, and a mile south of the town of Frisco. To get there, you have to hike up to the base of Mount Royal.

In 1866 General Buford discovered the Old Victoria Mine at the base of Mount Royal. General Buford worked on developing the mine, mining for gold and copper, and he built a very advanced and expensive mill. After this was completed, he left and no one in this area heard from him again. This was before Frisco was even a town.

In 1872, investors from Pennsylvania took control of the area, named it Masontown after their hometown, and built a reduction works plant at a cost of $75,000. They updated the existing mill and made a substantial investment into Masontown.

By 1880, the town had a population of about 300. Unfortunately, the mine just was not a good producer and never fulfilled expectations. But, the miners kept trying.

In 1912, an avalanche wiped out part of the area, and then in 1926 another avalanche wiped out the rest of the area. The town was completely abandoned.

During prohibition the area had a new life as a bootlegging operation, which lasted until the end of prohibition, when the site was abandoned for good.

PARKVILLE (Park City)
Summit County

Directions: Coming from Dillon, take State Highway 9 south for 10 ½ miles, until you reach Tiger Road, which follows the Swan River, and the Colorado Trail. Take Tiger Road for a little over six miles. At this point, just before the Swan River Forks, the left fork is the Middle Swan River, and the right fork is the South Swan River. Georgia Gulch will be to your right. Directly north at the mouth of the Georgia Gulch is the site of Parkville.

In August 1859, thousands of prospectors crossed Georgia Pass from the east, following the Colorado Trail, in search of gold. At the mouth of Georgia Gulch a town formed called Parkville. The gold rush was on! Rumors spread that men in Georgia Gulch were placer mining a pound of gold a day per man.

By 1860, the gulch was filled with men and the population soon peaked at 10,000. A post office was established 13 December 1861. The town became a supply center for the area. There was Whittmore's General Store, Kettle's Meat Market, The Chapin House Hotel, The Henry Weiss Brewery, Gallatin's Saddle and Harness Shop, Peabody and Company's General Store, and more businesses along Parkville's Main Street. There were two grand halls, and even a theatre with a forty-foot stage opened by Jack Langrishe and Dougherty, actors who gained fame from acting at Central City. The theatre featured such then famous actors as Lillian Gish and Frank Fay. Millions in gold were mined out of Georgia Gulch, reportedly $3 million in the first few years.

J.J. Conway and Company built a mint, which stamped out gold coins in $2.50, $5.00, and $10.00 denominations. The only coins in existence from this private mint are in the Smithsonian Museum. Mrs. Carl Modisett from Denver found the mint's original dies from 1861, in her attic, and she donated them to the Colorado Historical Society.

One of the early settlers of Parkville, Daniel Ellis Conner of Bardstown, Kentucky, kept a diary that was later published as *A Confederate in the Colorado Gold Fields*.

By 1863, mining in the area slowed, and the people left. Parkville was abandoned and buried under its own tailings piles.

PRESTON
Summit County

Directions: Coming from Dillon, take State Highway 9 for 10 ½ miles, until you reach Tiger Road on your left. Take Tiger Road for a little less than a mile until you reach Forest Route 300, which is also called Gold Run Gulch Road. Take this road to the south end of Gold Run Gulch. This is the site of Preston. Preston is three miles northeast of Breckenridge by trail or about six by road.

In 1875 rich gold strikes were found on the northern top of Gibson Hill. The prospectors working in the area formed the little town of Preston. The only businesses in town were a store, sawmill, boardinghouse, and a saloon. If the town residents needed anything else, it was only a short trip to Breckenridge. By 1882, *The Colorado Business Directory* reported the population at 150.

In 1884 the Jumbo Lode was discovered, turning out to be the richest gold mine in the area, supporting Preston through good times and bad times. Other good properties in the area included the Adelia, the Discovery Belt, the Inferno, the Intermediate, the Surles, and the Timothy. These mines contained gold, silver, copper, and lead. The Gold Run Mining Company, Consolidated Preston Mining Company, and Gold Run Placers were headquartered in Preston.

Gold production dwindled, and the market for silver crashed in 1893. Some men continued to mine into the 1930's, but the town was abandoned.

Traces remain of the many ditches and canals that had been dug for extensive hydraulic mining. These ditches and a few cabin ruins are all that remain of the town of Preston.

REXFORD
Summit County

Directions: Coming from Dillon, take State Highway 9 south for 10 ½ miles until you reach Tiger Road on your left. Take Tiger Road, which follows the Swan River and the Colorado Trail, for a little over five miles until you reach Forest Route 354. Bad weather has washed out this road, so you will need a four-wheel drive vehicle. Turn left on this road, following the North Fork of the Swan River for three miles, and the site of Rexford will be on your right, hugging the timberline, in a high alpine meadow.

Rexford was a small company town on the North Fork of the Swan River, formed after gold discoveries in 1880. The Rexford Mine Corp. was established in 1881 and owned and operated the area mines. The Rochester King, later named the Arastre King, was the best mine in the area. Daniel Patrick discovered the mine in 1880, and by 1881 the mine was producing $5,000 a month. Other good mines in the area were the Ben Franklin, the Black Swan, the Santa Claus, the Sweet William, and the White Swan. The corporation that operated these mines started with an investment capital of $100,000. Although it had other mining operations elsewhere in the state, it continued to pour money into the Rexford area mines. Rexford had a post office, but it was closed in 1883, when the town became essentially a company town. Then, the mail was brought to town from Montezuma two times a week. The town had a false-fronted hotel, a boarding house, a general store, a saloon, and a mill.

The production of the area mines declined so much that within a few years the town was abandoned. Only a few foundations can be seen today.

ROBINSON   (Camp Robinson, Ten Mile)
Summit County

Directions: Robinson is 294 miles from Denver at an elevation of 10,778 feet. Leaving Frisco on State Highway 91, the site of Robinson is 11 miles south, and on the right side of the road. The site of Kokomo is one mile north.

In the early spring of 1879, two prospectors, J. Sheddon, and Charles Jones, discovered rich carbonate ore on the eastern slope of Sheep Mountain. In the 1860's prospectors had engaged in extensive placer mining in the area, but the gold had been played out within a year, and the prospectors deserted the area.

In 1878, Sheddon and Jones had been grubstaked by George B. Robinson, a successful Leadville merchant, who provided supplies to prospectors in return for a 50% partnership on whatever the men would find. The prospectors were very lucky and discovered almost a dozen silver mines. Robinson immediately bought out his partners and was able to get financing from a couple of New York companies. He then started the Robinson Consolidated Mining Company. Prospectors, miners, and fortune hunters rushed to the area, and a town formed that they called Ten Mile City. With the money that poured in from his mines, Robinson developed the town. He built the Robinson Hotel, and someone else built the Bonanza Hotel. Robinson built a huge smelting and milling works in October 1880. The main part of the building measured 120 feet by 56 feet. Then, four sawmills were built, a school was built, then "Our Lady of the Snows" Catholic Church was built in 1882 under the close supervision of Father Fertaur. Soon many more businesses were built. In 1880, the D.& R. G. Railroad was built through town.

The name of the town was changed to Robinson to reflect George Robinson's efforts to build the town and make it a success. The town flourished, and Robinson, one of the newest Carbonate Kings, was very popular.

In November 1880, George B. Robinson was elected Lieutenant Governor of Colorado. Robinson was confident that he could successfully manage both his mines and the affairs of the state. In December 1880, Robinson and J.W. Jacque disputed the ownership of the Smuggler Mine. Robinson was the legitimate owner, but claim jumping was not uncommon. After hearing that a group of 100 claim jumpers were planning to grab his mine, he stationed armed guards around his property. Robinson was worried about his property and went to check on it. Failing to properly identify himself, he was shot in error by his own guard on 27 November 1880, just a month after being elected Lieutenant Governor of Colorado.

Robinson's accidental death was a crisis for the town, but the town continued to flourish until the late 1880's.

The Summit County Circular printed the following article on December 31, 1881:

MINING NEWS
THE ROBINSON

The New York Tribune has lately contained a number of dispatches from this point on the subject of the Robinson mine, which in attempting to depreciate the stock, have overstepped the bounds of truth. One dispatch dated December 20, says that, "there is little doubt the mine is exhausted." On the contrary there was on that day as there is to-day very grave doubt on the subject. It adds, same date, "Work on the east shaft was discontinued on Saturday." It disingemously neglects to explain that the discontinuance was caused by the necessity of changing the pump. It adds, same date, "Professor Ashburner`s assays were salted." Everybody knows he made none. It adds, same date, "The mine is shipping six carloads daily of ore of very low grade." We published in our last number the official statement that the week ending December 23, the shipments were eight loads per day. It adds, same date, "Heavy liabilities, other than those due the smelters, will mature in thirty and sixty days." The only heavy liability due by the Robinson on December 1, besides the overdrafts on Argo and La Plata, was the balance of $32,000 due the Jacque claimants, half of which has since been paid off, it adds same date, "The mine will probably shut in thirty days." There is no prospect whatever of anything of the kind. It is known that the ore in the eighth level is improving, and the managers will naturally find out the value of the ore in the ninth level before they think of shutting down. On this tissue of false statements the Mining Records properly remarks: "After having been used to spread broadcast over the country the belief that the Robinson property was one of almost exhausted richness, and thus enable certain speculators in New York, Denver, and Kokomo to rid themselves of the stock which they held, at from ten to thirteen dollars a share, that part of the *New York Tribune* ostensibly set apart for mining intelligence, is now being employed with amazing audacity, through such paragraphs as the above, to depreciate the probable value of the same property -- to the end of making the ebbing market in Robinson shares as profitable to the same men as was the flood-tide. The game however is so transparent that it ought to deceive no one. No one indeed should fail to detect an "exhausted mine" and closing ore shipments, which identify them as of the same paternity as that of the once daily extravagant dispatches to the *Tribune* from the same quarter, in regard to the reported exhaustless ore sources of the then marvelously rich Robinson mine.

The Summit County Circular
31 December 1881

The town had two newspapers, *The Robinson* and *The News*.

In 1882, *The Colorado Business Directory* listed the population at 500. By 1885, the population had doubled to 1,000 according to *Crofutt's Grip-Sack Guide*. The main mines in town were the Grace, and the Kelly, and the Robinson all owned by the Robinson Consolidated Mining Company. The Iron Mask, also one of the town's good mines, was named in *The Colorado Business Directory*.

The silver mines consisted of carbonites, sulphites, and galena. In 1893 when the big silver crash hit, most of the mines closed down. The town was almost deserted.

The Colorado State Directory for 1896 listed the following for Robinson:

ROBINSON
Mining town in Summit County, on the U. P. Ry., 16 miles northeast of Leadville. Population 100. Altitude 10,778.

Gallahger. D H assayer and drugs
Hamilton, W H, assayer
Hopkins, G G , assayer
Kutzleb, Wm, mayor

McDonald, S P, mining
Rice, N S, groceries & hdwre
Robinson Hotel, M Allen, prop
Walker, Wm, justice peace

By 1910, only 150 people lived in the town.

In 1920, Ten Mile Creek had flooded the town site, partially submerging Robinson's wooden sidewalks, and the foundations started sinking.

The state has erected a marker to commemorate the site of Robinson. Tailings piles left by the Climax Molybdenum Mining Company cover the site.

ST. JOHN (Coleyville, Sts. Johns)
Summit County

Directions: St. John has an elevation of 10,800 feet and is located about a mile south of Montezuma. Leaving the eastern edge of Dillon on U.S. Highway 6, go east, passing Keystone along the way, for six miles. At that point take Primary Forest Route 5, which will be on your right, for five miles. At this point, on the right will be Forest Route 275, which follows St. John's Creek, for about two miles. The road is not too bad, but a four-wheel drive vehicle is recommended.

St. John was a beautiful little mountain town at the base of Glacier Mountain, and about a mile south of Montezuma. In 1861 a party of miners was hunting in the area. After running out of bullets the men improvised, using as makeshift bullets the small pebbles they had picked up from the ground that they thought resembled lead. Three years later, one of the men was in Nevada, when he saw some high-grade silver ore. With a close look at the ore, he realized that the ore was the same as the small rocks that they had used as ammunition. The man told an acquaintance from Empire, John Cooley, that he thought there was high-grade silver ore on Glacier Mountain. John Cooley raced to Glacier Mountain and made one of the earliest silver discoveries in Colorado. He devised his own crude furnace, and melted his silver ore into bullion. The ruins of his furnace are about a half mile from the St. John Mine, nestled against the mountain.

By the next year in 1865, prospectors were everywhere, looking for a silver vein of their own. Bob Epsey, one of these prospectors, had spent the previous night drinking grain whiskey, and was feeling too sick to go anywhere that day. He rested under a tree, by some rock outcroppings. When he attempted to get up, he pulled himself up by grabbing a rock. The rock broke off in his hand, exposing high-grade silver ore.

By 1865, the Boston Silver Mining Association had just been formed.

By 1867, the little settlement was officially a town. Originally called Coleyville, the name changed by the end of 1867. It was officially called Sts. Johns, after St. John the Baptist and St. John the Evangelist. Later, the plural Sts. Johns was dropped to just St. John.

The Boston Silver Mining Association was changed to the Boston Silver Mining Company, and then to the Boston Mining Company in 1878.

The Colorado State Directory for 1878 listed the following for St. John:

ST. JOHN

Mining camp in Summit County. Population, 150. One store. St. John Smelting Works.

St. John was soon a company town run by Eastern capitalists from Boston. The town consisted of a three-story boarding house, a company guesthouse, a superintendent's house, which held the finest furniture that money could buy from Europe, a foreman's house, homes for miners with families, a company store, an assay office, and a few other buildings. The Bostonian capitalists established a library, imported foreign newspapers into the town, and kept the residents informed on current events. A very conservative group, they did not allow saloons or gambling in their town.

The men of St. John reportedly went a mile north to Montezuma to visit the saloons and brothels. They visited Montezuma whenever they had the opportunity.

Winters here were so brutal that the snow often completely covered the houses and the boarding house, and the men would have to use the second story windows as a door. By 1885, the mines' production slowed, and according to *Crofutt's Grip-Sack Guide of Colorado*, the population dropped to 40. The town was very isolated, and the silver was not as plentiful as it had been. Then, in 1893 the price of silver crashed, and the town of St. John was abandoned.

SWAN (Swan City)
Summit County

Directions: Swan is eight miles northeast of Breckenridge. From Dillon, take State Highway 9 south for 10 ½ miles until you reach Tiger Road, which follows the Swan River, and the Colorado Trail. Take Tiger Road for four miles, and on the right side of the road, at the mouth of Brown's Gulch is the site of Swan City. It is only three miles from Lincoln.

Swan was platted out on 11 May 1880. The post office was opened that summer. The Swan City School was established in February 1884. Placer mining was the main occupation of the men in town, although there was some quartz mining. The best ore in the area assayed at $800 a ton. Abundant game included elk, deer, bear, turkey, and grouse. Trout fishing was excellent.

The best mines in town were the Cashier Lode, the I.X.L. mines, and the Hamilton Lode. The following advertisement for the local stage line was printed in *The Breckenridge Daily Journal* on Saturday 18 June 1881:

DAVE`S BRONCHO STAGE LINE
BRECKENRIDGE, Colorado, February 18,1881
On and after February 21st, Dave`s Broncho Stage Line will leave the Grand Central hotel and the Wilderness postoffice at 8 a.m. for Swan City, Delaware flats and Galena gulch. For freight or passage apply at domestic bakery, 183 Main street, or of Dave and his Broncho wherever you can catch them.

The Breckenridge Daily Journal
Saturday, 18 June 1881

The Breckenridge Daily Journal printed the following article on Wednesday, 14 December 1881:

North Swan, Dec. 10, 1881
Another week has passed quickly and Saturday night is gladly welcomed by all, from superintendent to the newly promoted "wheeler," who bears his honors as meekly as if a lineal descendent of the Emerald Isle. The weather has been stormy and blustering and an additional foot of beautiful snow increase the depth of the winter mantle so profusely spread over the hills. Many of the

employees of the Rochester Queen, the principal mine, are also owners of prospects and lodes and their interests are materially served by the important developments being made, the future prospects are freely discussed and among other points two very necessary ones form the principal subjects. The importance of the camp requires the establishment of a postoffice, and that immediately. At present mail comes to Swan, six miles distance, three times a week, a daily mail to this point is imperatively needed and a petition is being circulated to accomplish that end. The establishment of this route is as much a matter of necessity to the merchants and business men of Breckenridge, where all the business of the camp is transacted, as to those living along the proposed route. The attention of the county commissioners is called to the necessity of improving the roads in the Swan river country. The building of two lines of railroad down the Blue has cut up the wagon road very badly, and when the spring opens it will be almost impassible. From Braddockville to Swan City, up the Swan, the placer mining enterprises are certainly ruining the wagon road, by the breaking of the ditches. Several small bridges are needed early in the spring to make it passable. From the Blue to this point a county roads has been laid out but not regularly recorded. From Swan to the center of the North Swan mining district a road has been made by private enterprise-principally by Alison, Miller, & Co. who extended last season upwards of $500 to make it passable. From this point to the head of the valley, to Akin & Brothers claims, is a distance one and a half miles. The prospective importance of this mining district requires immediate attention. The mining interests of South, Middle, and North Swan, will require not less than five hundred men during next season and the road if properly handled by the road overseer, will go far towards making all the road improvements necessary.

The Breckenridge Daily Journal
Wednesday, 14 December 1881

The Colorado Business Directory for 1882 listed the following;

SWAN

Mining camp in Summit County, 3 miles from Lincoln City and 8 from Breckenridge. Tri-weekly mails. Population in mining season, 300.

Allison, Miller & Co., mines.
Board & Dickson, saloon.
Eddy Mining Co.
I.X.L. Mining Co.
Loope, Geo. L.L., postmaster.
Loope, Jarchow & Co. gen'l mdse.

In 1885, *Croffut's Grip-Sack Guide of Colorado* reported that 100 people lived in Swan, and the post office was called Braddock.

In the 1890's the mining slowed. Perhaps it was the big dredges working the Swan River that put the placer miners out of business. The town was abandoned before 1900. Nothing is left but a few foundations and many tailings piles.

SWANDYKE
Summit County

Directions: Coming from Dillon, take State Highway 9 south for 10 ½ miles until you reach Tiger Road, which follows the Swan River, and the Colorado Trail for a little more than six miles. You should come upon Forest Route 6 on the left. This is a four-wheel drive road. Take this dirt road for about five miles, and the site of Swandyke is on the left, on the Middle Fork of the Swan River.

Swandyke was an isolated gold mining town that peaked in the 1890's. At an elevation of over 11,000 feet, Swandyke endured brutal winters. Because of its location, the town and the mines were often inaccessible in the winter, and a lot of the residents would leave for the season. The town was just west of the Continental Divide, facing steep mountainsides that were soaring walls of rock. The town was divided into an Upper Swandyke and a Lower Swandyke, with at least a mile between the two sides.

In the 1890's the population peaked at 500, with regular mail service. Stagecoaches traveled to and from Breckenridge, 14 miles southwest. Stagecoach service was also traversed Georgia Pass and on to the town of Jefferson, which was 17 miles southeast.

The miners bragged that the ore here was richer than anywhere else in the district. The Swandyke Gold Mining and Milling Company operated several area properties, including the Three Kings Mine. The White Swan Mining Company also operated properties in the area. Most of the mines were above the timberline. Swandyke's main street was lined with all of the normal businesses of a town. It had stores, restaurants, saloons, blacksmith shop, butcher shop, and barber shop. One of the hotels was The Summit House. Cabins were scattered around the town.

After the year 1900, the mines became less productive. This dwindling productivity, combined with the extremely bad weather in the winter and the town's isolation, contributed to the town's death.

Today, all that remains of Swandyke are a few ruins and foundations.

SWANVILLE
Summit County

Directions: Swanville is eight miles northeast of Breckenridge. Coming from Dillon, take State Highway 9 south for 10½ miles until you reach Tiger Road, which follows the Swan River, and the Colorado Trail. Take Tiger Road for a little more than six miles, and then you will come upon Forest Route 6 on the left, which is a four-wheel drive road. Take this road for about 1½ miles. The road will fork. The left fork will take you to Swandyke, and the right fork crosses the creek to a meadow. This meadow and a little beyond is the site of Swanville.

Swanville was an early placer mining town, which died out early, leaving little trace of its existence.

TIGER
Summit County

Directions: Coming from Dillon, take State Highway 9 south for about 10 ½ miles until you reach Tiger Road on your left. Take Tiger Road, which follows the Swan River, and the Colorado Trail, for about 4 ½ miles, and the site of Tiger will be on your left.

Tiger was a little company town just north and on the Swan River. It was four miles east of Gold Run Gulch.

In August 1859, prospectors came from Southpark. They crossed the Continental Divide, explored Gold Run Gulch and Swan River, and discovered gold in the sands. Rumor said that some of the placer claims were so rich that the men would pan out, or sluice out, a pound every day for every man. The Weaver Brothers recovered about 96 pounds of pure gold within theOR first six weeks of sluicing in Gold Run.

In the mid 1860's after the discovery of the Tiger Lode, a little town formed. After that the Royal Tiger Mines Corporation was incorporated, it bought and combined the area mines, such as Swan City's old I.X.L. Lodes, and the Cashier Lodes, both listed in *The Colorado Mining Directory* for 1888. The town of Tiger had running water, steam heat, and electricity. The Royal Tiger Mines Corporation also supplied the town with a general store, a large bunkhouse, cabins for those miners who had families, and a doctor's office.

Starting in 1898 with the Blue River, by Breckenridge, which was 8 ½ miles southwest, dredges were used in the area. Gold dredges had been used up and down the Swan River. The ruins of one dredge can be seen today. The last dredge was shut down in 1939. The town had a fascinating Ball Mill, which had a huge hopper that held iron balls, each about the size of a small bowling ball. These iron balls were used to crush ore to a fine powder so that the gold could be extracted. In the summer the ore would be hauled out by wagon to the railroad, and in the winter sleds were used.

The town of Tiger held on longer than most of the other mining towns in the area. In 1918, a deadly influenza epidemic spread death throughout the town, striking down the weak and elderly. The town was abandoned in 1939, when the Royal Tiger Mines Corp. shut down their operations here. In 1973 a fire destroyed all of the town buildings. Nothing is left.

ALTMAN (Midway)
Teller County

Directions: Altman is three miles northeast of Victor. Altman was built between Gull Hill and Bull Cliff in what is called a saddle. The easiest way to get there is to go to Goldfield. From Goldfield you can see Altman. Roads all over the mountainside lead to it, making Altman an easy town to reach.

Altman was formed around 1894 when its post office was established. Two years later in 1896, the town of Altman was incorporated. Sam Altman was the operator and owner of one of the district's first sawmills and stamp mill. His sawmill created steam, and he used that for his stamp mill. The town Altman was named after him. The town residents were union miners, and the leader of these men was King Calderwood. It was said that back in Scotland he had worked in a mine since he was nine years old. He took good care of the union miners. The mine owners would try to get away with paying their miners as little as possible. In 1894, the union went on strike. The miners built a fort on Bull Cliff and waited to go to war with the mine owners. There was a fight and shootings. Martial law was declared in the district. At least 1,500 men fought each other, with deaths on both sides. The town was so violent the undertaker offered group discounts on the weekends.

Altman was a busy little town. Its population peaked at 2,000 in 1900. It has an altitude at 10,700 feet. The town did not have water, but residents built a pipe system and piped water from two miles away out of Grassy Gulch. There were about ten saloons; the Mint, The Monte Carlo, the Silver Dollar, and the Thirst Parlor were the most popular. There were a couple of restaurants, at least four boarding houses, a drugstore, and two general stores. The town covered the top of Bull Hill.

The Pharmacist mine is also on top of the hill. Two pharmacists from Colorado Springs came to town to try mining, knowing nothing of the subject or how to find gold. They decided to just throw their hat up and file their claim where it landed. Well, as soon as they started digging, they struck it rich. They found half a million dollars in gold.

The Buena Vista and the Victor were also good mines, but the American Eagle was the biggest and best.

In Altman's boom years, they needed strong, courageous lawmen. Sheriff Mike McKinnon tried to protect the town and faced down six armed troublemakers from Texas. The sheriff was shot and killed, but not before shooting and killing the six outlaw Texans.

After Sheriff McKinnon's death, Marshal Jack Kelly was charged with protecting the town. "General" Jack Smith swooped down upon the town. He was the leader of the "Smith Gang" that had harassed the Cripple Creek area for years. One night, Smith and his gang got drunk, took over the town, and released the prisoners from the jail. Marshal Kelly arrested Smith. Smith then posted bail and spread word that he would be waiting for the marshal at the saloon. When Marshall Kelly walked into the saloon, Smith shot first. He missed the marshal, but the marshal's fatal return fire found its mark.

In 1899, the mayor of Altman was Thomas Ferroll. He liked the town the way it was and wanted to prevent it from being modernized by a telephone line run through town. When the telephone company started to set up the telephone poles, Ferroll went up to the foreman and told him to consider himself arrested. He told the 19 workmen to go home, and he took an axe and cut down a few telephone poles. The Colorado Telephone Company challenged the mayor, the mayor backed down the next day, and the telephone line was built.

A fire ruined much of the town in 1903. There are still some buildings on the main street. At one end of the town was the City Hall and at the other lower end is the Union Hall foundation. There is a beautiful view of Pike's Peak.

When Altman was at its peak in 1900, it overgrew the hilltop where it had been platted. It spilled toward the north and absorbed the town of Midway. When Altman's residents left, so did residents of Midway.

It is rumored that union activists Harry Orchard and Steve Adams planned their terrorist activities here.

Right around the hill from here is Squaw Gulch and at the top of this gulch is their Glory Hole. It is a 150-foot cave in which the miners actually climbed the steep rock sides with ropes to excavate the site for gold.

No one lives in Altman now. There are only a few buildings, foundations, and memories.

ANACONDA (Mound City, Squaw Gulch, and Barry)
Teller County

Directions: Anaconda is three miles southeast of Cripple Creek on State Highway 67, almost halfway between Victor and Cripple Creek.

In 1894 the town was founded. Before it was called Anaconda, the miners called it Squaw Gulch, because they had unearthed the decomposed body of an Indian woman while digging for gold. The miners later called the town Anaconda after the nearby Anaconda Mine. The most productive mines in the area were the Anaconda, the Doctor, the Jackpot, and the Mary McKinney. The Mary McKinney Mine produced over $11 million in gold, but today it is only a great tailings pile at the mouth of the gulch. In its day, Anaconda had about 2,000 residents and was the biggest town in the Cripple Creek area. Photographs from the time show a long main street dotted with a row of buildings stretching almost the length of the gulch. If you look closely at these old photographs, you can discern hundreds of houses. A railroad spanned almost the length of the gulch.

Mound City was a mining camp next to Anaconda that had grown up around the Brodie and Rosebud Mills. Barry, named for Horace Barry, was next to Anaconda. Horace Barry had found some of the best claims in the area, but he lost them in poker games and bar tabs. As Anaconda grew, it absorbed the community of Barry.

In 1904, a fire started in a meat market. A fierce wind caught the flames, and within minutes the entire gulch was engulfed in a roaring fire. The town was destroyed. Nothing remained. Even the telephone poles were consumed. Neighboring Cripple Creek and Victor tried to help, dispatching their volunteer fire departments to help fight the blaze, but it was too late. The residents never rebuilt their community.

Looking at the valley now, it is difficult to believe this was ever the site of a thriving city.

CAMERON (Gassy, Grassy)
Teller County

Directions: Cameron is six miles north of Victor on County Road 122.

In the beginning, the Cameron area was a little ranching town, called Gassy after a resident with a horrible gastrointestinal problem. In 1892, the railroad came through town, and a little company platted out a city here, calling it Grassy after the meadow. But soon the railroad quit coming through town and most of the people moved away.

In 1899, an investment company bought the town site, renamed it Cameron and set up their new town. They built a 30-acre park on the south side of town. It included a stadium that could seat 1,000 people, a dance pavilion, zoo, picnic ground, playground, and some restaurants. It was a beautiful park. People from neighboring towns would come on the weekends and holidays. They named it Pinnacle Park. On 4 September 1900, the town threw a Labor Day party, and 9,000 people showed up, ready to celebrate and have fun. The park charged a 10-cent admission. The school was up to date, and the city buildings were well maintained. The population peaked at this time at around 500. A few good mines supported the town: the Elsmere, the Isabella, the Lansing, the Wild Horse, and a few others.

In 1903 labor wars started erupting, and then the mines in the area dried up. The people moved away.

Nothing is left. If you want a beautiful trip with a four-wheel drive, follow the old railroad grade road that leaves Cameron and rambles north to Colorado Springs.

ELKTON (Beacon Hill, Eclipse, Arequa)
Teller County

Directions: Elkton is on State Road 67 between Cripple Creek and Victor on State Road 67. Elkton is two miles north of Victor.

Elkton was founded by a prospector new to the Cripple Creek area, who staked his claim here. He named it Elkton after some old elk antlers on the ground. The prospector was short of money and food, so he traded grocers in the next town a half interest in his claim for $36.50 worth of groceries.

The Elkton mine did very well. It produced about $15 million in gold. Another mine just on the other side of town, the Cresson mine, was founded around 1900. Shortly afterwards, the owner died and left the property to his son. The property had not been worked for a few years, and the young man who owned the property now asked a geologist to see if the property was any good. The geologist said it was good property. In 1915, the Cresson Vug was discovered. The miners had dug a shaft 1,200 feet deep and struck a pocket of pure gold, 20 feet deep and 30 feet wide. The owner installed a vault door at the opening of the mine and hired armed guards had to protect shipments. The Cresson mine produced over $51 million in gold.

Elkton was a bustling little town of around 3,000 people in 1905. It had lots of saloons, hotels, and a few grocery stores. Elkton absorbed the nearby mining camps of Arequa, Beacon Hill, and Eclipse. When the mines quit producing, the residents moved away. There is quite a bit of the town left. There might even be one or two old miners around the town looking for something that might have been overlooked.

The Colorado State Business Directory for 1896 listed the following for Elkton:

ELKTON
Camp in the Cripple Creek mining district, in El Paso county. Location Elkton, Raven and other mines. Population 300. See also Anaconda.

Cowen, Wm, genl mdse
De La Vergne, E M, pres and mgr Raven Gold Mining Co
Eldorado Mining Co, C B Seldemridge, pres, E M De La Verne mgr
Elkton Consolidated Mining & Milling Co. Geo Bernard pres, J H

Avery secy-tres W N Bainbridge mgr
Elkton House, Miss E D Eddy, prop
Ellison, O S, grocs & meats
Ingham Consolidated Min Co
Keener, G L, mgr Nugget M & M Co
McMurtie, M C, postmaster
McMurtie, T A groceries
Raven Gold Mining Co, E M De La Vergne, pres and mgr
Requa G & S Mining Co. Thos S Craven, pres, E M De LaVerne mgr
Richardson, C L, groc, meats
Robinson, Thos N, broker
Rogers, Mrs. Mary, boarding
Warner, J E, assayer

GILLETT
Teller County

Directions: Gillett is about seven miles northeast of Cripple Creek off State Road 67.

Gillett was the gateway to the Cripple Creek gold district, setting by itself to the north on level land, with beautiful views and scattered waterfalls. The population peaked at about 1,000. There were a few churches, a school, a casino, boarding houses, and a horse track. A little bit of mining was done around town. The area was pockmarked with prospect holes.

The Lincoln was the best mine in the area. Because of litigation, the owners shut down the mine, but it still has substantial ore reserves. Midland railroad had a huge railroad terminal here. The level ground was perfect for it. The city had its own water system supplied by snowmelt from Pikes Peak. They had fire hydrants in case there was a fire. A little power plant was built so the town could have electric lights this was a family town. There was just enough money to take care of their families. In the summer of 1895, Joe Wolfe, nicknamed "Arizona Charlie," and his associates planned a huge three-day bullfight event from August 24 to 26th. Famous matadors and huge bulls were brought in from Mexico. Advertising all over the state and beyond, tickets were sold for $5 each. When the day of the big event arrived, 50,000 people showed up to the Gillett race track, where a bull fighting arena had been built. The crowd had traveled far, and they wanted a good show. When the time came for the actual bullfight, the bulls did nothing but stand there. The crowd started to get angry. The matadors started to torture the bulls to make them fight. They still would not fight. The bulls just sat there. The crowd started to riot. The matadors tortured the bulls, eventually killing them. The show was over after a couple hours. "Arizona Charlie" and his associates had to hide out in the Gillett jail until the angry crowd broke up and returned to their homes. The only good thing that came from this incident is that the slaughtered bulls were given to the poor people of Gillett for food. In 1911, when the racetrack was no longer used, a farmer plowed it under and planted oats.

The citizens of Gillett started to leave in the early 1900's. The mining in the area was no longer productive; the area was no longer used as a railroad terminal. Gillett is now a field, with the ruins of a church, quite a few homes and remnants of a boomtown jail.

GOLDFIELD AND INDEPENDENCE
Teller County

Directions: Goldfield is one mile northeast of Victor on 122. Independence is one mile past Goldfield. These two towns grew so close to each other that they eventually became one town, with Goldfield absorbing Independence.

In 1894 two prospectors discovered gold. A plumber, James Burns, and a carpenter, James Doyle, staked a claim on Battle Mountain. They started to work their claim and ran low on money. They knew nothing about prospecting. They added John Harnon as a third partner for his financial help. Before these men had dug ten feet, they struck rich gold. The men called it the Portland Mine. They wanted to keep it a secret, so they worked at night, stashing their ore. Soon other miners found out, and they tried to file claims on this property. The men organized the Portland Mining Company and, luckily for them, they had stashed a lot of ore before other miners found out about their strike. The Portland Mining Company successfully fought all of the lawsuits and claims. Later, the company's assets were worth $100 million. The Portland Mining Company then formed the town of Goldfield in 1895 at the bottom of Battle Mountain. The site had previously been a cow pasture, a level beautiful field surrounded by rolling hills at the base of the mountain. In 1895, one mile north of Goldfield, the town of Independence sprang up.

As the two towns grew and flourished, they became one. And the population peaked at over 3,000. Goldfield and Independence were very family oriented and union based. *The Goldfield Leader* and *The Goldfield Times* were the two local newspapers. There were four churches, five hotels, a lot of saloons and stores, and a couple of schools. Three railroads had terminals here, and they ran through the town. This was the third biggest city in the Cripple Creek area.

In 1903 when the Cripple Creek area was rocked by the labor wars, violence also struck this little town. Union leaders put all their workers on strike in 1903, which in turn caused the mine owners to bring in workers to work in place of the striking miners. The first group of strikebreakers, called "scabs" by the union miners, started working, and there was trouble. Harry Orchard planned a big explosion in the Vindicator Mine, the fourth largest gold producer in the area, producing $27 million to date. It was November 1903, and Orchard had hoped to blow up a bunch of strikebreakers. Harry went in the mine and set the explosives. Instead of strikebreakers, he blew up and killed a shift boss and a mine superintendent. Harry had failed at his attempt at the mine, so he started planning for his next union terrorist attack. The Hull City Mines and the Vindicator Mine supported the town of Independence. In June 1904, Harry Orchard set explosives at the Independence railroad station. The resulting huge explosion blew up the whole station and killed many men. One source says 19 men were killed; another source reports 13. Many more men were seriously injured. Orchard, who was a member of Western Federation of Miners, helped end the labor wars by getting everyone's attention. He tried to blow up Colorado's Governor James H. Peabody, Supreme Court Justice William Gabbert, and Justice Luther Goddard. Orchard was unsuccessful. He just wanted to help the miners who worked so hard to support their families. Goldfield and Independence were family towns. Owners of these mines were multimillionaires who paid their miners $2.00 an hour with no benefits. Sometime after this, Orchard set explosions that killed Idaho's former governor. He was arrested and spent the remainder of his life in Idaho's State Prison.

Independence and Goldfield are now both near-ghost towns with perhaps a handful of residents.

STRATTON (Winfield)
Teller County

Directions: Coming from Cripple Creek, take the road that goes almost directly east from the middle of town by the museum. It will go over Gold Hill. Take that road east for three miles and there is Stratton / Winfield. Midway is a half mile away.

Or, coming from Victor, take the Range View Road, which leaves town from the northeast corner of Victor. Take this road five miles north. You will take a road to the left at Globe Hill, and the town is on the left side of the road after about a fifth of a mile.

The town of Stratton, also known as Winfield, was built by Winfield Scott Stratton, Cripple Creek's first millionaire. Coming to Colorado in 1872, he was a carpenter by trade, and that is how he supported himself at first. Stratton prospected on the side. In 1881, he lived in Silver Cliff, in 1882 he lived in Red Cliff, in 1883 he lived in Aspen, in 1884 he lived in Tin Cup, and he always spent his winters in Colorado Springs. In 1885, Stratton still had not found any gold, so he took an assaying and mineral course at Colorado College. It was the spring of 1891. Stratton had just spent his winter in Colorado Springs, and it was time for him to try prospecting. He left for the Cripple Creek area. He could have taken the stage, but he walked to save money. He had his two burros with him. Stopping briefly to talk with and visit with Bob Womack, who was prospecting in the hills, Stratton then walked to Battle Mountain. All of the other prospectors were miles away from here by Bull Hill. Stratton noticed a granite outcropping at the bottom of Battle Mountain. It looked like an ordinary granite ledge, but Stratton had a feeling. Taking samples from the rock, he tested them himself using a blowpipe. The test showed a little gold float, but not what he was seeking. Stratton prospected elsewhere, but he kept coming back. Finally on the Fourth of July in 1891, Stratton returned to his spot and staked out two claims, calling one The Independence and the other, The Martha Washington. After digging out just a few shovels of dirt, he struck a very rich vein of gold. The Independence would eventually produce $25 million in gold. Stratton developed his mines and was worth millions. Stratton had lived for 20 years in poverty, living on carpenter jobs and prospecting. To the shock of all the social elite, instead of buying a mansion like the rest of the Bonanza Kings, he bought a two-story house that he had built years before, bordering on the business district of Cripple Creek. Stratton then bought bicycles for all the laundry women in town so they could ride to work.

Stratton was the only person to offer financial help to Horace Tabor when he was in need and shunned by his former friends. And, after Horace Tabor's death he saved the Matchless Mine for Baby Doe.

John Harnon was an out-of-work, unsuccessful prospector who wandered into Cripple Creek looking for work. Harnon was down on his luck. Stratton met Harnon in town and put him to work immediately as a cook. After a few months of getting back on his feet and saving some money, Harnon went prospecting up Battle Mountain, where he met James Doyle and James Burns who had staked their claim there and were unsuccessfully working it. When Harnon became the third partner, his experience helped them make their big strike, the Portland Mine.

In 1899, Winfield Scott Stratton sold his Independence Mine to a company in England for $11 million.

In 1902, Winfield Scott Stratton died, leaving most of his fortune to establish the Myron Stratton Home, an institution on the edge of town to care for residents who were physically unable to support and care for themselves, whether a young orphan, the handicapped, or an elderly person who has outlived family. The world needs more Winfield Scott Strattons.

The town built by Stratton was his headquarters for the Independence Mine, and it was his company town, made up of half brick structures and half wood frame structures. The citizens here were mostly miners living in the many boarding houses.

Sadly, little is left of the town of Stratton. A mining company came in the 1980's and tore down many of the buildings.

REFERENCES

Newspapers
Animas Forks Pioneer
Apex Pine Cone, The
Bonanza News, The
Boulder News, The
Carbonate Weekly Chronicle, The
Caribou Post, The
Crystal River Current, The
Daily Budget, The
Breckenridge Daily Journal, The
Denver Republican, The
*Elk Mountain Bonanza, The (*renamed *The Gothic Miner)*
Goldfield Leader, The
Goldfield Times, The
Gothic Miner, The
Gothic Record, The
Hillerton Occident, The
Leadville Chronicle, The
Marble Times, The
Maysville Chronicle, The
Miners' Record, The
Ouray Plaindealer, The
Pilot, The
Pueblo Chieftan
Red Mountain Journal, The
Robinson, The
[Robinson] News, The
Rosita Index
Rocky Mountain News, The
St. Elmo Mountaineer, The
*St. Elmo Rustler, The (*renamed *The St. Elmo Mountaineer)*
San Juan Expositor
Silver Lance, The
Solid Muldoon
South Arkansas Miner, The
Summit County Circular, The
Summit County Times, The
Summitville Nugget, The
Ten Mile News
True Fissure, The

Ward Miner, The
White Pine Cone, The

Autobiographies
Dyer, John Lewis. *The Snow-shoe Itinerant*. Cincinnati, 1890. (This was reprinted in 1975.)

Ellis, Anne Heister Flemming. *The Life of an Ordinary Woman*. Boston 1929.

Mclean, Evalyn Walsh. *Father Struck it Rich*. Boston, 1936.

Other
Crofutt, Geo. A. Crofutt's Grip-Sack Guide of Colorado. Omaha, 1885. (This was first published in 1881, with an enlarged edition published in 1885. The rare 1885 edition was reprinted in 1996 in Boulder by Johnson Books.)

Colorado Business Directory, various years.

U.S. Federal Census for the State of Colorado, various years.

Willison, George F. *Here They Dug the Gold*. New York, Third Edition. 1946.

INDEX

94 (Ninety-Four) 61 82
94 (Ninety-Four) Mine 61
A.Y. Mine 161
ABBEY, C.F. SMELTER 130
ABBEYVILLE 130 131
ABBOTT Mining Co. 57
ADA Mine 63
ADAMS, Arthur 17 Charles D. 237
 Ed. 21 Geo. S. 90 J.H. 127 Steve
 295 Thad 17 21
ADELAIDE 156
ADELAIDE Mine 97 156 162
ADELAIDE PARK 97
ADELIA Mine 280
ADELINE Mine 130
ADERHOLD, Mr. and Mrs. Fred
 234
ADIT Mine 21
ADLER, J.S. 127
AGASSIZ 176
AIRLINE Tunnel 13
AJAX Mine 147
ALBANY Mine 75
ALICE 61 62 63 64 82
ALICE Mine 63 64
ALLEN, M. 285
ALLIE BELL Mine 35
ALLISON, MILLER & CO. 290
ALLNOTT, L.S. 23
ALLUM, Mr. 5 6 7 8
ALMA 203 204 205
ALMA STATION 203 218
ALPHA Mine 191
ALPINE 25 26 27 28 29
ALPINE Hotel 28
ALPINE Lode 25
ALPINE Loop 150
ALPINE STATION 113
ALPINE Tunnel 35 113
ALPINE and VIRGINIA CITY
 Toll Road. 131
ALTA 248

ALTA Mine 248
ALTMAN 294 295
ALTMAN Pass Road 47
AMAZON Hill 131
AMERICAN CITY 101 102 109
AMERICAN EAGLE Mine 294
AMERICAN FLAG Mine 4 107
AMERICAN HOUSE 258
AMERICAN UNION Mine 276
AMES 249
AMETHYST Mine 187 188
AMITY 178
AMITY Mine 178
ANACONDA 296
ANACONDA Mine 296
ANDER, George 262
ANDERSON, Charles 67 Chas 229
 Eli 78 Fred 67 J.R. 72
ANDREWS, John 211
ANIMAS FORKS 145 146 152 153
 154 233 234 235 236 237 238
 239 240 245
ANKLAND Mine 58
ANNA Mine 28
ANNIE MASCOT Mine 102
APEX 101 102 103 104 109
APEX Hotel 103
APPERSON, Wm. 91
ARASTRE KING Mine 281
AREQUA Camp 298
ARGENTINE 255 256 257
ARGENTINE Pass 81 255 259
ARGENTUM 154
ARGO Mine 228
ARMSTRONG, --- 74 J.D. 66 Lon
 67
ARNOLD, P.R. 273
ASHBAUGH, A. 107
ASHLAND Mine 185
ASPEN 37 47
ATCHINSON, SPONSOLIER
 and WHALEY 141

INDEX

ATLAS Mine 200
ATWOOD, --- 128
AUSTIN, --- 55 E. 90 H.E. 91
AVERY, J.H. 298
AVERY'S SMELTER 124
AZTEC Mine 224
-B-
BABCOCK 30
BABCOCK, Colonel 30
BACHELDER, William 67
BACHELOR 187 188 189 190 191
BACHELOR CITY Hotel 189
BACHELOR Mine 148 187 188 191
BACHELOR Mountain 187
BACHUS Mine 98
BACKMAN, S.J. 90
BADGLEY, C.W. 63
BAINBRIDGE, W.N. 299
BAKER, C.A. 66 Charles 91 Daniel 88 Col. S.H. 251
BALANT, Ralph 104
BALCH, J.B. 23 Nellie 23
BALD Mountain 107
BALDWIN 114 121
BALDWIN & MACKAY 229
BALDWIN & SAVARD 49
BALLINGER, W. 214
BALTIMORE 105
BANKER Mine 59
BANNER GROUP Mining Co. 66
BARACOTA Mine 58
BARLOW & SANDERSON 127
BARNETT, H.M. 188
BARNIE & RICE 188
BARRET, W.E. 90
BARROW, St. James 199
BARRY 296
BARRY, Horace 296
BARTHEL & BURCHART 125
BARTLETT, A. 91
BASSHAM 54
BASSICK, Edmund C. 86

BASSICK CITY 86
BASSICK Mine 86 87 92
BASSICKVILLE 86
BATTLE Mountain 301 303
BAY, Geo. W. S. 189
BEACON HILL Camp 298
BEAR Mountain 117
BEARDSLEY & McNEAL 125
BEATTY, Ed 234 P.B. 246
BEAUMONT & CO 90
BEAVER CITY 31 60
BECKER, Emil 55
BELCHER, A.C. 189
BELFORD Falls 57
BELL, J.H. 246
BELMONT and TONOPAH Mining Co. 248
BELMONT and WAGNER 248
BELLE of the WEST Mine 98
BEN BUTLER Mine 245
BEN FRANKLIN Mine 281
BENGLY, Geo. 269 270
BENSON, Israel 22 Martin 273
BENT, C.H. 153
BERG, Charles 67
BERGER Bros. 12
BERNARD, Geo. 298
BERTA, Joseph "Peanuts" 114
BERTIE, Robert 252
BESHORE & GRAHAM 14
BIEDELL, Mark 228
BIG FIVE CORP. 17 19 22
BILES, J.A. 189
BILL YOUNG Mine 245 247
BILLINGS, C.L. 78
BINDER, Robert 19
BINGHAM, Dave 177
BISMARK Mine 276
BIXBY, A. 2
BLACK, Sam 269
BLACK EAGLE Mine 118
BLACK HAWK 1 8 22 101 102 107

INDEX

109 112
BLACK HAWK Mine 27 28
BLACK PRINCE Mine 156
BLACK QUEEN Mine 118
BLACK SWAN Mine 281
BLACK WONDER Consol. Mining Co. 153 Mine 152
BLAIR, W.H. 18
BLAKE, O.F.
BLANDIN, A. 214
BLATCHFORD, G.E. 43
BLATCHLEY, F.C. 91
BLOODSWORTH, Officer 168
BLOSSOM, Charles F. 90
BLOUCETT, Moses 90
BLUMBERG, A.D. 195
BLUSH, Daniel 91
BLUSH & BAKER 91
BOARD & DICKSON 290
BOBBITT, John 211
BOHN, John 83
BONANZA 227 228 229
BONANZA Hotel 282
BONANZA KING Mine 148
BONANZA Mine 228
BONANZA Peak 228
BONE, John 68 John Jr. 67 John Sr. 67
BONHOMME Mine 154
BONITA 227
BONITO 65 66
BONNIE, Sue 173 174
BOOK, Miss 18 21
BOOMERANG Mine 241
BOSTON OCCIDENTAL Mining and Milling Co 101
BOSTON [SILVER] Mining Assoc. / Co. 286
BOUGHER 71
BOUGHTOWN 159 162 176
BOULD Mine 180
BOULDER 13

BOULDER COUNTY Mine 11
BOWERMAN 115 116
BOWERMAN, J.C. 115 116 Mrs. J.C. 115
BOYD, Col. 88 89 Mr. 34
BOYD and CO. 269
BOYLE 48
BRADDOCK 262
BRADDOCK, Dave 262
BRAUN, Theo. F. 90
BREED & CUTTER 2 6 9 11
BREWSTER, J.W. 90 91
BRICKENSTEN Mines 27
BRICKNER A.J. 256
BRISCOE BLOCK Hotel 47
BRITTENSTEIN Hotel 28
BRITTLE SILVER Mine 260
BROADE, E.E. 189
BROMLEY 157
BROMLEY STATION 157
BROOKLY & ALEXANDER 270
BROSS, Gov. 255 Stephen Decatur – see Decatur, Stephen
BROWN, --- 212 E.M. 237 E.S. 237 Ed. M. 234 F.W. 103 Geo. 107 Jasper 88 Leander 72 Policeman 268 Richard 72
BROWN & CAMP 238
BROWN & CO. 125
BROWN'S Gulch 288
BROWN'S Spring 88
BROWNE, Gen. Sam 208
BRUBAKER, H.E. 72
BRUCE, Fletcher 125
BRUM, August 107
BRUMLEY 157 158
BRUNER, Mrs. Rosa 189
BRUNNER, Mrs. Lydia C. 147
BRUNOT AGREEMENT 150
BRUSH, F.W. 49
BRYANT, Joe 273
BUCHANAN, Jas. 230

INDEX

BUCK HORN 18
BUCKLES, M.A. 125
BUCKSKIN JOE 160 203 204 205 206 207 208 207 264
BUCKSKIN Gulch 204
BUCKTOWN 179
BUELL, Jas. 90
BUENA VISTA Mine 294
BUFFALO Flats 262
BUFORD, General 278
BULL CLIFF 294
BULL DOMINGO Mine 120
BULLDOZER Mine 185
BULLION KING Mine 134
BULLION SMELTER 65
BUNT, John 67
BURKLEY Mining Co. 118
BURNS, --- 74 Fred C. 72 James 301 304
BURRETT, J.B. 187
BURROWS, Capt. Abe W. 236 245 Charles 145
BURROWS Lode 107 245
BURROWS PARK 145 154 233 234
BUSH, William ("Bill") H. 169
BYERS, William 58
BYRON, E.E. 273

-C-

CALDERWOOD, King 294
CALDWELL, E.K. 83
CALIFORNIA Gulch 33 159 161 168 172 175 176 203 206 207 208 264
CALIFORNIA Mine 107
CALHOUN, J.H. 246
CALLEN, A.W. ("Old Grizzly") 62 63 Jim S. 62
CALLERY Bros. 76
CALUMET 55
CALUMET Iron Quarry 54
CALUMET Mine 141

CALVIN, C.L. 187
CALWELL & BLEWITT 125
CAMBELL Bros. 28
CAMBELL & CARR 195
CAMERON 297
CAMP, W.S. 91
CAMP BIRD Mine 156 161 200 202 237
CAMP FRANCES 17
CAMP FULFORD 94
CAMP GROVE Mine 107
CAMP HALE 99
CAMPIAN, John 182 183
CAMPIAN Hotel 183
CANICE Mine 192
CAPITOL CITY 146 147
CAPITOL CITY Mine 146
CARBONATE Camp 162
CARBONATE Mine 161
CARBONATEVILLE 258
CARBONATEVILLE House 258
CARDINAL 11 15
CAREY, Mr. 239
CARIBOU 1 – 16
CARIBOU Flats 1 14 15
CARIBOU Hill 1 4 5 9 15 16
CARIBOU House 4 13
CARIBOU Mine 1 2 6 10 13 14 261
CARIBOU Stage Co. 3
CARIBOU Tunnel 13
CARLSTROM, Chas. 234 238
CARNEY, Thomas 126 127
CARR, John E. 195
CARSON 148 149
CARSON, Christopher 148
CARSON Camp 148
CARTER, Dennis 162 M.B. 20
CASE Mine 261
CASHE CREEK 32 33 34 159 183 264
CASHIER Lode 288 293
CASS & CO. 206

INDEX

CASS, EATON & CO 205
CAT GULCH 54
CATALPA Mine 118 162
CATLIN, L.C. 72
CAUFIELD, W.J. 176
CAVE Mine 97
CENTENNIAL Mining Co. 86
CENTRAL CITY 12 13 82 102 103 107 108 167 208
CHAFFEE, Jerome 164
CHAFFEE ACT 71
CHALK CREEK Ranch 181 Toll Road 47
CHAMPIAN & TANDEY 221
CHAMPION Mill 155 Mine 154
CHANCE 230 231
CHANDLER Mine 148
CHAPIN HOUSE Hotel 279
CHAPMAN, Mr. 26
CHARLESTON Mine 161
CHARLETON, Mrs. W. 189
CHAVANNE 71
CHENEY, Edw. F. 82 M.A. 23
CHERRY Smelting Co. 118
CHICAGO & LONDON Tunnel Co. 59
CHIHUAHUA 259 Hotel 259
CHINN CITY 78
CHLORIDE HILL Mine 75
CHRISHOLM, Malcolm 78
CHRISTIANSON, Wilborn 34
CHRYSOLITE Mine/Tunnel 27 28 48
CHURCH and COOK 276
CINCINNATI Mine 276
CINNAMON Pass 154 233 245
CLANCY, James 78
CLARENCE 136
CLARK, --- 247 Calvin 276 George 276 M.J. 225
CLARK & GRUBER 205
CLARK, MORT & CO. 185

CLASBY, Watt 139
CLAUS, Fred 67
CLEAR CREEK Canyon 78
CLEAR CREEK Gulch 57 58
CLEMENS, C. H. 28
CLEOPATRA Mine 118 188
CLIFTON Hotel 48 49 House 273
CLIMAX 267
CLIMAX Mine 165 Molybdenum Mining Co. 258 267 273 285
CLINTON Mine 28 152
COCHETOPA Placer and Gold Quartz Mining Co. 127
COFFIN, TRAVER & CO. 153
COLCORD, A.B. 273
COL SELLERS Mine 273
COLE, George 1 George A. 73
COLEYVILLE 286
COLLIER, Phil 176
COLLINS, C.M. 199
COLORADO MINERAL BELT Mining Co. 49
COLORADO Mountain 101
COLUMBIA Mine 19 21 253
COLUMBUS Mine 240
COLVILLE, John B. 189
COMBS, Robert 108
COMER, D.B. 108
COMIER, W.S. 72 73
COMMODORE Mine 188
COMMONS, Patrick 195
COMMONWEALTH Mine 28
COMO 209 214 220
COMSTOCK Mine 98 184 185
CONGER 261
CONGER, Mine 1
CONGER, Sam 1 11 261
CONGRESS Mine 197
CONKLIN, John 21
CONNER, Daniel Ellis 279
CONQUEROR Mining and Milling Co. 76

INDEX

CONSOLIDATED PRESTON Mining Co. 280
CONTINENTAL Mine 28
CONVERSE, Geo. A. 103
CONWAY, J.J. 279
COOK, Gen. 212 J.E. 73 T.P. 108
COOLEY, John 286
COOMBS, Fred G. 234 235
COOPER 71 128
COPPER Creek 123
COPPER GLEN 194 195
COPPER Gulch 227
CORAL Mine 11
CORNWALL 223
CORNWALL, John 223
CORNWALL Silver Mining Co. 223
CORSAIR Mine 191
CORY, G.W. 153
COSGROVE, John 13
COTTONWOOD Camp 232
COTTONWOOD Pass 37 Toll Road 37
COVODE Mountains 76
COWEN, Wm. 298
COWLES, Brothers 5 Clinton 76
COX, George A. 73 John E. 67
COZE, Geo. 67
CRACKER JACK Mine 154
CRAIG, Frank 72 73
CRAVEN, Thos. S. 299
CRAWFORD, E.H.
CREAGER, J.C. 91
CREEDE 142 188
CREEDE, Nicholas 42 187
CRESCENT Mine 162
CRESCO Mine 148
CRESSON [VUG] Mine 298
CRILLY & CO. 259
CRIPPLE CREEK 21 172 295 296 298 300 301 303 304
CRISS, J.D. 49

CRIST, J.H. 234
CROESUS Mine 13
CROFUTT, George A. 40
CROMPTON, W.G. 82
CROOKE Bros. 150
CROPSEY, Mrs. Allen 91
CROSBY, J.E. 78
CROSBY Furnace Workings 76
CROW, Vernon 213
CRYSTAL 117 118 119 143
CRYSTAL Mill 117
CRYSTAL Mountain 117
CUBA Mine 241
CUMBERLAND Mine 184
CUNNINGHAM & SLOAN 189
CURTIS & BARRETT 214

-D-

D & R.G. Mining Co. 59
DACOTAH Mine 245
DAILEY, John 78 Patrick 78
DAILEYVILLE 78
DALTON Lode 118
DANA Mine 161
DANIELS, R.N. 91 William P. 19
DANIELS & ROBERT 91
DAVEY, John A. 67
DAVIES, Dan 270
DAVIS, --- 28 Mrs. 250 Cyrus W. 243 E.L. 250 G.W. 153 J.R. 14 John C. 73 Rev. William 192
DAYTON 8
DE LA VERGNE, E.M. 298 299
DEAN, L.R. 270
DEATHRIDGE, H. 152
DECATUR 89 255 259
DECATUR, Stephen 89 255
DECATUR House 256
DEE Creek 212
DEFIANCE Mine 228
DEFRANCE, Fred 67
DEL MONTE Mine 188
DEL NORTE 224 225 228 236

INDEX

DELAMAR 85
DELAWARE CITY 262
DELAWARE Flats 262 263
DELAWARE Mine 256 Mining Co. 256
DENTRESS, C. 82
DENVER CITY Mine 75 165
DEW DROP Mine 21
DEWITT, Henry 76
DEXTER, Jim 165 166
DEXTER Mine 165
DIAMOND Lode 191
DIANTHE Mine 261
DIEHLE & CO. 90
DILLON, Pat 162 Richard 162
DINNING, James 91
DISCOVERY BELT Mine 280
DOCTOR Mine 296
DODGE, John 67
DODGE AND CO. 141
DODSON, H.R. 14
DOE, Baby – see TABOR, Baby Doe
DOHERTY, J.H. 189
DOLOMITE Mine
DONACASTER, Mayor 268
DONNELLY, Leo 3 13 14
DORAN 215
DORCHESTER 120
DOUGHERTY, --- 279
DOWD, J.W. 273
DOWNEY, Father 188
DOYLE, James 78 301 304 W.A. 67
DUBLOND, --- 271
DUCHAT, Ed. 28
DUDLY & STEEL 124
DUGGAN, Mart 168
DULONEY, David 78
DUMONT, John 66
DUNBAR, T. 214
DUNCAN, Miss 17 18 20 21 Mr. and Mrs. 234 J.C. 90 William 237
DUNCAN & FRANZ 90
DUNCAN Camp 232
DUNCAN Mine 165
DUNCOFF, Thomas 78
DUNDERBERG Mine 148
DURBIN Bros. 98
DYER, Father John Lewis 48 207 264 265 266
DYERSVILLE 264 265 266
DYES, Albert 234
DYKE Mine 107

-E-

EADES, A.B. 189
EAGLE 96 141
EAGLE Mine 141
EARLY BIRD Mine 240
EAST LEADVILLE 215
EAST PORTAL 106 111
EATON, G.C. 117
ECKART, Mrs. 234
ECLIPSE Camp 298
ECLIPSE Mine 42 171 172
EDDY, Miss E.D. 299 Edward 169
EDDY Mining Co. 290
EDWARDS, Bros. 91 Mr. 247
EGYPTIAN QUEEN Mine 184
EICHELBERGHER, E.C. 230
EKKARD, Mrs. 238
EL DORADO Mine 75 Mining Co. 298
ELDORA 23 24
ELEPHANT Mine 276
ELITHORPE, Bob 225
ELLISON, O.S. 299
ELK Creek 212
ELK Mine 165
ELK MOUNTAIN Toll Road 47
ELKO 126 143 144
ELKTON 298 299
ELKTON Consolidated Mining & Milling Co. 298 Mine 298

INDEX

ELKTON House 299
ELLIOT, John 82
ELLIS, Anne 228 Pasco 108
ELLSBURG, P.N. 145
ELSMERE Mine 297
EMPIRE 76
EMPIRE Mine 165
EMPRESS JOSEPHINE Mine 228
   Mining Co. 229
ENGINEER Pass 233 245
ENTERPRISE Mine 120 197
EPSEY, Bob 286
ERIE Con. Mining and Smelting
   Co. 40
ESMOND, John 224
ESMOND Mine 224
ESPINOSA Brothers 203
ESTER, John 92
EUREKA, 154 233 234 240 241
   241 245
EUREKA Mine 156
EVAN, J.G. 42
EVANS, D.T. 83 Gov. John 58 266
   R.W. 28 S.J. 252 Walter 252
EVENING STAR Mine 28 162
EVERETTE 158 196
EVERETTE, C.M. 158
EVERETTE HOUSE 158
EVERGREEN Mine 102
EXCHEQUER Mine 227 228
EXCHEQUERVILLE 227
EYRE & LATHROP 256
EYSSELL, Hugo 83

-F-

FABIAN, J.F. 92
FAIRPLAY 162 210 211 212 213
   216
FAIRPLAY DIGGINGS 210
FALCO, Prof. 27
FALL RIVER 64
FANCY PASS 99
FARLEY Mine 118

FARMER, Joseph 83
FARNCOMB, George 276 Harry
   274 276
FARQUHER, John 98
FAUL and CLARK 25
FEATHERS, Amasa 40
FEELEY, Martin 73
FERGESON, --- 160 Wm. 237
FERROLL, Thomas 295
FERTAUR, Father 282
FETCHER, E.C. 230
FILERS, Anton
FINLEY, W.F. 185
FISH, Chas. 103
FISHER, --- 133 Geo. R. Thos 92
   Mrs. Thos. 92
FITCH, Chas R. 28
FITZMORSE, Thos. 73
FITZPATRICK, F.J. 230
FLACK Mine 107
FLAG STAFF Hotel 238
FLANCIER Mine 75
FLICK and STRAND 141
FLORA BELL Mine 35
FLORENCE 85
FLORESTA 121 122
FLORESTA COAL Mine 122
FOELL, Mr. and Mrs. 234
FOOSEL Camp 30
FOOTE, Phil. 267 268
FOREST HILL Mine 120
FOREST HOUSE Hotel 246
FOREST QUEEN Mine 133 134
   Mining Co. 135
FORT Garland 203 227
FORT Massachusetts 83
FOUR METALS CO. 248
FRANCES 1 – 22
FRANK, William 251
FRANKLIN, G.W. 152 153
FRANKLIN Mine 261
FREE COINAGE Hotel 188

INDEX

FREELAND 65 66 67 68 69 70
FREELAND Consolidated Mill 67 80 Mine 66
FREEMAN, A.D. 152 Edgar 76 J.H. 152
FREEMONT Pass 181
FREMONT, Col. John C. 150 204 220
FRENCH Gulch 274 275
FRENCH Pass 275
FRENCH Peak 98
FREY, R. 98
FRIDDICH, Jacob 78
FRISBIE, Mr. 207
FROST & CO. 260
FRYER, George 162 164
FRYER HILL 162 164 172
FULFORD 94 95 96 97
FULFORD, Arthur H. 95 96 James 96
FULLER, F.W. 135

-G-

GABBERT, William 302
GABLE, William 73
GABLEMAN H. & CO. 103
GALENA 85
GALENA Gulch 138
GALLAGHER, --- 161 Senator 156
GALLAHGER, D.H. 284
GALLATIN, --- 279
GARDNER Mine 107
GARMAN, J.C. 18
GARSTIN, C.H. 126
GASSY 297
GASTON, Rev. 192
GATES, John S. Theodore 73
GENESSEE Mine 192 Vandervilt Mining Co. 199
GEORGE III Mine 148
GEORGE WASHINGTON Mine 152
GEORGIA Gulch 279

GEORGIA Pass 275 279
GILBERT, J.G. 67
GILL, A.E. 20
GILLETT 300
GILLETT, F. 188
GILLIAM, Charles 90
GILPIN, Gov. 214
GLACIER Mountain 286
GLADSTONE 243 244
GLASS, --- 5 Geo. 20
GLEN COVE Mining Co. 57
GODDARD, Luther 302
GOE, Bill 104
GOLD ANCHOR Mine 82
GOLD BRICK District 142
GOLD BUG Mine 54
GOLD Creek 141 142
GOLD KING Mine 184 243 244 249 248
GOLD LINKS Mining Co. 142
GOLD PARK 98
GOLD PARK Mining Co. 99
GOLD PRINCE Mine 240
GOLD RUN Camp 32
GOLD RUN Gulch 280 293
GOLD RUN Mining Co. 280
GOLD RUN Placers 280
GOLDFIELD 294 301 302
GOLDRING, 2 4
GOOD HOPE Consol. M & S. 256
GOODWIN, Charles 150
GORDON Claim 183
GORMAN, M. 13
GORMAND, HUDSON & SIEBER 91
GOTHIC 123 124 125 126 127 128 129 131 143
GOTHIC Hotel 125
GOVERNOR KING Mine 276
GOVERNOR Mine 200
GOW, Wm. 82
GOYNE, I.D. 67

INDEX

GRABALL Camp 210
GRACE Mine 284
GRAHAM, Major 89 Tim 83
GRAND ISLAND, 23
GRAND VIEW Hotel 91
GRAND VIEW Mine 4 13
GRANDE MESA & ROCK CREEK Co. 118
GRANITE 31 32 57 59
GRANT, H.H. 83 Bros. 230 Ulysses S. 12 125 134 143 228
GRANT Hotel 125
GRASSY 297
GRAY, John 92
GRAY'S Peak 259
GRAYBACK Gulch 83
GREAT EASTERN Mine 146
GREAT NORTHERN Hotel 179
GREAT REPUBLIC Mine 13
GREENFELL, J.H. 108
GREENHOW, Thomas 73
GREENLEAF, E.B. 236 238
GREEN'S GULCH Camp 30
GREGORY Hotel 54
GRIFFIN, Superintendent 269 270
GRIFFIN Mine 178
GRIZZLY Gulch 47
GRIZZLY Mine 28
GROW & SMITH 222
GUGGENHEIM, Simon 162
GUIONNEAU, Benj. 147 Bros. 147
GUNNISON 32 39 114 131 132 140 144 230
GUNSTROM, John 73
GUSTON 192 193
GUSTON Mine 192 194
GUT, Robert 272

-H-

HAENNI, John 49
HAGER & CO. 91
HAINES, H.T. 17
HALF MOON Gulch 157 168
HALF-WAY House (Note: there is more than one) 126 128 158 181 219
HALL, A.L. 237 J.M. 255
HAMILTON 214 220
HAMILTON, A. 84 A.K. 84 Ayres 83 Bros. 84 Earl 214 W.H. 284
HAMILTON Lode 288
HANCOCK 30 35 36 45 53 113
HANCOCK Placer Claim 35
HANDCART Gulch 213 Pass 221
HARDER, J.N. 269 270 272
HARDING, Mr. 21 George 78 Jerry 20
HARDSCRABBLE Mining District 88
HARGER, Rev. 21
HARNON, John 301 304
HARPER, Lewis 67 Stephen 73
HARRINGTON, Dennis 73
HARRIS, --- 247 Fred 234 Hart 205 207
HARRIS & COMPANY 118
HARRISON, Charley 210 Edwin 28 Senator 28
HARRISON Mine 28 118
HARRISON Tunnel Co. 48
HARTER, S.B. 13
HARTVILLE 43
HARTZ Camp 30
HARVARD CITY 37
HATCHET, --- 74
HUDLER, Clark 230
HAVENS & BOTTLESON 11
HAVERLY 134
HAWK, Mrs. Loretta G. 82 Mose 5
HAWKE, William 67
HAWKEYE Mine 75
HAWKS, A.M. 82
HAWTHRON & ROOT, 185
HAYNES, J. Bruce 206
HAZELTON, E.H. 28

INDEX

HEACK, John 43
HEAD, John 67
HECK, Casper 256
HEILSCHER, August 189
HEIMBOLT, Fred G. 234
HEITLER, R. 271
HENDRICKSON, H.A. 19
HENNEBERG, Misses 189
HENRIETTA Mine 165
HENRY, A.E. 199
HENSON 150 151
HENSON, Henry 150
HERMAN House 13
HERMAN, G.P. 23
HERZINGER & HARTER 13
HESS, Jacob 141
HESSEY, W.W. 49
HESSIE 23 24
HIBERNIA Mine 165
HICKMAN, Henry 185
HICKOCK, Wild Bill 134
HICKS, J.M. 73 J.R. 108 Thos. 176
HIDDEN TREASURE Mine 150 151 191 200
HIGGENBOTTOM, Joseph ("Buckskin Joe") 205
HIGH MUCH-A-MUCK Mine 147
HIGHLINE Mine 261
HILL, Major 253 Professor Nathaniel 1
HILL and BUTLER 269 270
HILL TOP 217
HILL TOP Mine 217
HILLER, Edward 130
HILLERTON 130 131 132
HIMROD, Peter 71 72 74
HINES & DOWD 269
HINKLE, P.R. 20
HINKLEY, W.S. 135
HINMAN, Mrs. S.E. 14
HIVELY Mine 28
HIX, Donald 104 Horace 104

HOBBS, W. 124
HOCKING, Joseph 108
HOGE, J.M. 91
HOLLIMAN, Tom 211
HOLLINGSWORTH, Judge 236
HOLMAN, Jack 22
HOLMAN & MORRISON 117
HOLMES, W.H. 91
HOLT, E.T. 84
HOLY CROSS CITY 98 99
HOMESTAKE Creek 98
HOMESTAKE Mine 13
HOMESTAKE Mountain 98
HONEYCOMB Mine 205
HOOK, George 162 163
HOPKINS, Dr. 4 10 G.G. 284
HOPKINS & CO. 90
HORSESHOE 215 217
HORSESHOE Mountain 215
HORTENSE 38
HORTENSE Mine 28
HOTEL DEL MONTE 101
HOWARD, E.P. 234
HOWARDSVILLE 154
HOYT, --- 247 Fay 128 V.B. 91
HUBERT Mine 107
HUDSON & FISH Mine [Tunnel] 28
HUDSON & SIEBER 91
HUFFMAN, Mr. 17
HUGGINS, Al 267 268 Chas 14
HUGHES, W.M. 189
HUGHES' HOUSE Hotel 40
HULL CITY Mines 302
HUMBOLT Hotel 91
HUMBOLT Mine 88 89 200
HUNKIDORI Mine 98
HUNT, --- 247 John R. 234
HUNT['S] PLACER 247
HUNTER, A.V. 166
HURLEY, Pat 48 49 Williams 78
HUSS & SMITH 269

INDEX

HUSTED, H. 92
HUTCHINSON, Georgie 140
HYMAN, M. 269 270

-I-

I.X.L. Mines 288 290 293
IDAHO Mine/Tunnel 2 5 11 13 184
IDAHO Mountain 104
IDARADO Mining Co. 250 254
ILSE 85
IMOGENE Pass 249
INCAS Mine 147
INDEPENDENCE 302
INDEPENDENCE Mine 303 304
INDEPENDENCE Pass 47 [Toll] Road 157 158 182
INDIANA Gulch 261 264
INDIANA Mine 107
INEZ Mine 118
INFERNO Mine 280
INGHAM Consol. Mining Co. 299
INGRAHAM, Fred 92
IRIS 230 231
IRISH, D.C. 73
IRON CHEST Mine 27
IRON CITY 39
IRON CITY Power Plant 39
IRON CROSS Mine 58
IRON MASK Mine 148 284
IRON SILVER Mine 161
IRONTON 192 194 195
IRONTON Hotel 194
IRWIN 88 121 133 134 135
IRWIN, Dick 88 191 Rev. John 189
IRWIN City Water Works Co. 135
IRWIN Milling & Power Co. 135
ISABEL Mine 185
ISABELLA Mine 297
ITALIAN MOUNTAIN District 120

-J-

JACK'S CABIN 131
JACKSON 28 179
JACKSON'S DIGGINGS 80
JACQUE, J.W. 282
JAMES, Mr. 118 Jesse 168 William H. 169
JAPAN Mine 253
JASPER 223
JEFFERSON 216 291
JEFFERSON, Wm. 195
JENKERSON, Mr. 9 10
JENKINS, J.W. 189
JENNINGS, James 123
JENNINGS Camp 30
JERSEY CITY Mine 102
JESSIE Mine 27
JENNY LIND Gulch 105
JOHN Lode 247
JOHNSON, A.C. 78 A.J. 67 Mrs. A.J. 68 Alfred 73 Bros. 117 118 J.E. 273 Victor 67
JONES, Charles 282
JONES Mine 107
JOPLIN House 258
JORDIE, John 14
JOSEPH, Ed. 251 Howell 67 John 68 John D. 67
JUDD, Garwood N. 128
JULIA CITY Camp 232
JUMBO Lode 280
"JUNGLE" 253

-K-

KANE, Tom 234
KANSAS CITY Smelting Co. 28
KANSAS Mine 107
KEARNEY, S.D. 230
KEARNS, John 73
KEARSING 7
KEENER, G.L. 299
KELLER, Joseph 91
KELLEY, Brothers 136 J.P. 256 R.R. 135
KELLY, John 176 Marshal Jack 295
KELLY'S DIGGINGS 176

INDEX

KELLY Mine 284
KEMP, S.W. 103
KENNEDY, G.W. 17 Mrs. William 92
KENNEY, Jas. 229
KENT & CALLAHAN 91
KERBER, Captain Charles 227
KERBER CITY 227
KERKETT, George 22
KETTLE, --- 279
KILLDUIF, Thos. 270
KIMBER, --- 247 Charles E. 234
KING, John 176
KING COLE Mine 209
KINNEY, Willis 243 244
KINSMAN, Mrs. S.R. 18 21
KIPP & MORRIS 199
KIRWIN, J.J. 230
KIT CARSON Mine 148
KITTIE B. Mine 97
KITTY CLYDE Mine 205
KLAUBER, S.J. 130
KLOCK, H.F. 248
KNAPP, Charles 152
KNIGHT, Tom 211 W.L. 91
KOBY Bros. 108
KOESTLE, Jacob 238
KOHN, James 90 91
KOKOMO 255 258 267 – 273 282 283
KREUTZER Mine 191
KURTZ, Orion 92
KUTZLEB, Wm. 284

-L-

LA FOUNTAIN, E. 273
LA PLATA CITY 184
LA PLATA Mine 184
LADY ELENOR Mine 184
LAKE, M.C. 82 Professor 17
LAKE'S Camp 138
LAKE CITY 146 147 150 152 153 233 236 237 245

LAKE VIEW Mine 27
LAKEMAN, T.L. 152
LALLA Mine 61
LAMARTINE 70 7172 73 74 75 79
LAMARTINE Mine 69
LAMB, J.W. 82
LANG, John 73
LANGLEY, Rev. H.C. 90
LANGRISHE, Jack 279
LANNING, O.E. 73
LANSING Mine 297
LARIMER, W.L. 158
LAST CHANCE Mine 134 217
LATHROP, E.S. 256
LATTA, Lafe 20
LATTER, Thos. 271 272
LAURETTE 206
LAWSON 77
LAYTON Mine 97
LE VETA Pass 83 84
LEAD KING Mine 134
LEAD QUEEN Mine 119
LEADVILLE 26 27 28 32 37 63 98 124 131 134 138 159-174 175 176 178 179 180 181 182 209 219 227 258 267 268 269 273 282 284
LEAHY 28
LEASURE, Thomas 59
LEAVENWORTH Gulch 5
LEAVENWORTH Mountain 5
LEAVICK 217
LEAVICK, Felix 217
LEE, Abe 32 159 161 C.C. 189 George 146
LEGAL TENDER Mine 148 Stamp Mill 141
LEITER, Levi 161
LEOBELT, W. 214
LERIAN, Adam 59
LESHER & BURNAM 4
LEVI & LEWIN 190

319

INDEX

LIBERTY 232
LIBBY, Mrs. E. 68 Richard 67
LIENTHAL, H.H. 214
LIKINS, F.L. 189
LILLIAN & ANTIOCH Mining Cos.177
LINCOLN [City] 263 274 275 276 277 288 290
LINCOLN Mine 300
LINDGREEN, G.W. 271
LITTLE ANNIE Mine 224 Mining Co. 223
LITTLE BOB Mine 180
LITTLE CHARM Mine 42
LITTLE CHIEF Mine 162
LITTLE EARL Mine 130
LITTLE ELLEN Mine 180
LITTLE MATTIE Mine 28
LITTLE NELL Mine 48
LITTLE PITTSBURGH Mine 163 164 181
LITTLE PRINCE Mine 63
LITTLE VIRGINIA Mine 28
LITTLE WILLIE Mine 165
LIVERNASH, Adolphus 13
LIVINGSTON, David 86
LLOYD, Joseph 13
LOCKWOOD, G.D. 234
LOFTUS, J.P. 19
LOGAN, W.R. 49
LOGUE, W.O. 3 4
LOITUS, Professor 17
LONDON JUNCTION 218
LONE TREE Mine 66 Mining 67
LONGWELL, J.M. 269 270 271
LOOPE, Geo. L.L. 290
LOOPE, JARCHOW & CO. 290
LORD, Miss Alice 236
LOST CANYON 32
LOST LAKE 23
LOST TRAIL Mine 148
LOST TREASURE Mine 75

LOUCKS, W. 189
LOUISE Mine 180
LOVEL, "Chicken" Bill 162 164
LOWRY, Thos. H. 108
LUCHA, William 78
LUCKY Camp 232
LUDWIG, Henry 234
LUESLAY, R.C. 238
LUIS MARIA BACA GRANT #4 232
LUMLEY, David 67
LUNDBERG, A. 73
LUNDK, G.G. 73 P.G. 73
LUNDY Saloon 188
LYNCH, Geo. 189
LYON, O.P. 195
LYTLE, George 1

-M-

MACHEBOEUF, Bishop 48
MACKAY, John W. 66
MACKAY 444 Mine 110
MACKENZIE, John C. 187
MACKEY, Dick
MACKEY Mine 102
MacLEAN, A. 189
MACHMER, --- 239
MADONNA Mine 42
MAID of CARSON Mine 148
MAIER, John 20
MALLON, H.P. 189
MALTA 175
MAMMOTH 111
MAMMOTH Mine 11 28 118
MAMMOTH Tunnel Co. 48
MANIHULIN, P.J. 23
MANION, Mrs. M. 189
MANN, William J. 14
MANSFIELD 128
MARBLE 117 136 137
MARKLEY, Tailor 189
MARRA, John O. 84

INDEX

MARRILL, Wiley 73
MARSHALL, G.G. 73 Graham, 73 James Y. 73 Gen. William 40
MARSHALL Pass 40
MARTHA WASHINGTON Mine 303
MARTIN, Mr. 6 L.B. 141 William 1
MARY McKINNEY Mine 296
MARY MURPHY Mine 28 45 46 47 48
MASON, D. W. 230 Senator William 55
MASONTOWN 278
MASTODON Mine 245 246
MATCHLESS Mine 172 173 304
MAUDE S. Mine 247
MAY, I.S. 189 T.J. 127
MAY-MAZEPPA Mine 138
MAY QUEEN Mine 165
MAYDAY Mine 184
MAYER Mine 224
MAYSVILLE 40 41 42 51
McCALE, J. 270 L.E. 270
McCAMMON, Hugh 1
McCLANCY 128
McCLELLAN, Billy 207
McCLOUD, --- 239
McCLURE, J.E. 28
McCONNELL, Z. 234
McCOOK, Gov. 4
McCOURT, Elizabeth – See TABOR, Baby Doe
McCUNE, J.A. 270
McDONALD, A.K. 189 Dan 189 G.T. 273 S.P. 285
McDONOUGH, Bros. 230 John 78
McELHINNEY, A.M. 91
McENANY, Mrs. P. 236
McENTEE Bros. 235
McFARLAND, Mr. and Mrs. 234 Mr. and Mrs. Capt. 234 D.L. 73
McGONIGAL, Daniel 108

McGUIRE, John 73
McILVAIN, Harvey 153
McINTYRE, Charles H. 245
McKEE, Col. 210
McKENZIE, D.B. 189 Duncan 68 John A. 68 R.W. 68
McKINNIE Mine 241
McKINNON, Jas. 92 Sheriff Mike 295
McLAUGHLIN, H. 78 Wm. 90
McLEOD, R.J. 189
McMILLIAN, George 78
McMURTIE, M.C. 299 T.A. 299
McNEAL & REYNOLDS 5
McNEFF 91
McPHEE, Angus 189
MEAD, Albert 150
MEARS, Otto 194 197 237 241
MECHANICS Mining Co.. 57
MEDILL 71
MELROSE Gold Mining Co. 87
MELVIN, James A. 90
MELVIN House 91
MENDOTA Mine 97
MERCER COUNTY Mine 107
MERRIAM, George D. 28
MERRIAN, Capt. 38
MERRILL, N.C. 19 S.B. 17 19 20 S.L. 49
MEYER, Gustav 109
MIDDLE Mountain 59
MIDDLETON, 154 233
MIKADO Mine 156
MILLER, --- 42 Arnold 82 Charles L. 68 D.D. 91 F.L. 91
MILLINGTON & MOODY 223
MILLINGTON, Wm. V. 223
MINERAL Hill 230
MINERAL POINT 117 233 234 236 237 239 245 246 247
MINERAL POINT Mining and Reduction Co. 246

INDEX

MINERAL POINTS RED CLOUD Mine 233
MINERS Hotel 185
MINNETONKA Mine 28
MINNIE LEE Mine 152
MINNIE Mine 161
MINT Saloon 294
MISCHLER, Samuel 1
MISER Mine 223
MISSOURI Mine 107
MISSOURI VALLEY Mine 13
MITCHEL and FINE 136
MITCHELL, D.F. 271
MOFFAT, David H. 4 106 111 164
MOFFAT Tunnel 105 106
MOLLY Mines 98
MONARCH 42 43
MONARCH Mine 42
MONARCH Pass Toll Road 40
MONDAY, A.C. 189
MONITOR Mine 13
MONNIE MUSK Mine 75
MONSTER Mine 152
MONTE CARLO Mine 134
MONTE CARLO Saloon 294
MONTEZUMA 260 281 286 287
MONTEZUMA Silver Mining Co. 221
MOORE, D.H. 43 H.N. 67
MOORE, SCHWEEDER & LIVERNASH 11
MOORE & JONES 128
MORGAN, 3 John E. 68
MORNING GLORY Mine 156
MORNING STAR Mine 5 11 13 147 162
MORO Mine 147 156
MORRIS, S.L. & CO. 269 270
MORRISON, Alex 78, Judge 270 W.K. 194
MORROW, R.W. 98

MORTON, G.W. 84
MORTON House 84
MOSHER, W.H. 73
MOSQUITO 219
MOSQUITO Gulch 218 219
MOSQUITO Pass 206 209 218 219
MOSS, Capt. John 185
MOSS Mining Co. 67
MOUND CITY 296
MOUNT ANTERO 30
MOUNT CARBON 121
MOUNT CHAMPION 157 183
MOUNT CHAMPION Mine 157
MOUNT ELBERT 183
MOUNT McCLELLAN 81
MOUNT PRINCETON HOT SPRINGS 38
MOUNT ROYAL 278
MOUNT SNEFFELS 200 201
MOUNT VERNON Lode 11
MOUNT WHITNEY 98
MOUNTAIN BOY GULCH Mine 157
MOUNTAIN CHIEF Mine 30
MOUNTAIN CITY 107
MOUNTAIN House 273
MOUNTAIN VIEW Mine 152 Mining Co. 153
MOUNTZ 102
MULLEN, Joel K. 150
MUNROE Mine 27
MURPHY, J.J. 13 Jos. J. 13
MURRAY, --- 247 Dr. 236
MYERS, John 92
MYRON STRATTON HOME 304
-N-
NANKERVIS, Wm. 108
NANKERVIS & RHEM 108
NAPOLEON Mine 145
NATHROP 25 54
NATIONAL Mill 6

INDEX

NATIONAL BELL Mine 196
NATIVE SILVER Mine 13
NEDERLAND 13 15
NEELY & SANDERS 90
NEGLECTED Mine 184
NEGUS PLACER Mining Co. 76
NELSON, C.F. 187 Charles 90 E.E. 79 Olaf 243
NELSON Mine 191
NESBIT, Dave G. 23 Laura G. 23 W.G. 23
NEVADA CITY 107
NEVADAVILLE 107 108 169
NEW DISCOVERY Mine 162 164
NEW ENGLAND Hotel 130
NEW HOPE Mine 152 Mining Co. 153
NEW ORO CITY 162 176
NEW YORK CABINS 97
NEW YORK / LAST CHANCE Mine 187 188 189
NEW YORK Mine 97
NEW YORK Mountain 96
NEW YORK Smelting Co. 229
NEWARK 27
NEWARK CITY Mine 261
NEWELL, Samuel 14
NEWPORT 250
NI-WOT Mine 21 22
NICHOLLS, W.H. 68
NICHOLS, Wm. 108
NO-NAME Mine 2 13
NOBLE Bros. 108
NODINE, G. 91
NOLAN 95
NOLAN'S CREEK Camp 94 95
NORTH EMPIRE 76
NORTH LONDON Mine 218
NORTH STAR 138 139 140
NORTH STAR Mine 138 139
NORWOOD Mine 13
NOTMORE, C.W. 230
NUGGET 101 102 109 110
NUGGET Mine 109 Mining and Milling Co. 299
NUNN, L.L. 249
-O-
O'BRYAN, J.W. 63
O'CONNOR, City Marshal 168
O'SHEA, T.E. 73
OAKLAND HOUSE 258
OCEAN WAVE Mine 147
OFFENBACHER, W.A. 90
OFFICERS' BAR 83
OHIO CITY 141 142
OHIO Cons. Tunnel Co. 59
OHIO Mining Co. 141
OHIO Pass 121
OLD LOUT Mine 245 246
OLD VICTORIA Mine 278
OLDS Hotel 125
OLLIE, 79
OLNEY, Henry 130
ONEIDA Mine 145
ORCHARD, Harry 295 301 302
ORDWAY, W.F. 185
OREGON CREEK Camp 32
ORO CITY 156 159 162 176 177
ORPHAN BOY Mine 192 205 256
ORTH, G.S. 246
OSGOOD, K.C. 136
OURAY 201 202 237 245 246 251
OYLER, T.J. 4 5
-P-
PAGAND, W.H. 91
PAGE, J.B. 189
PAIGE CITY 274
PAINE, Thos. 273
PALMER, John 68
PANDORA 250
PANDORA House 250
PANDORA Lode 250
PANGBORN, Asa 225
PANGBURN & CUTHBURT 273

INDEX

PARK, Mrs. M.E. 98
PARK CITY 219
PARK Hotel 219
PARKINSON, J.B. 103
PARKVILLE 279
PARLIN, David 68
PARROT, Tiburcio 185
PARROT CITY 184 185 186
PARSONS, I.M. 108 John 220
PAT MURPHY Mine 46
PATRICK, Daniel 281
PAYMASTER Mine 40 192
PAYNE, B.D. 90 91 92
PEABODY, --- 279 Gov. James H. 253 302
PEABODY & CROSIER 214
PEABODY Mine 5
PEASE, Mayor J.G. 126
PEASE & HALL 131
PELICAN Mine 98
PELTON, J.A. 57
PENNSYLVANIA Mine 255 256 259
PENNSYLVANIA Reduction Works 91
PERKINS, Marc. G. 234
PERKINS Hotel 276
PERRY Lode 223
PERUVIAN ("PERU") Mine 256 259
PETERS PLACER Mine 241
PETTIBONE Mine 27
PHARMACIST Mine 294
PHILLIPS, John A. 73 William M. 73
PHILLIPS Lode 207
PICKEL, John H. 1
PIE PLANT Mine 120
PIERCE, John 13
PIERSON BROS. & SCRIBNER 238
PIERSON, C. and CO. 238

PIKE, Zebulon 204 220
PINE CREEK Hotel 103
PINE CREEK Mining District 102 109
PINE Mine 161
PINTO, Fr. 90
PIPER, J.P. 230
PLACER 83
PLACER Creek 83
PLACERVILLE 251 252
PLANDETT, T.F. 273
PLANTERS House 13
PLATTE CITY 210
POCAHONTAS Mine 88 89
POHNEAR, John 68
POLAR STAR Mine 97 147
POLLOCK, Mrs. Laura J. 189
POMEROY Gulch 47
POMEROY Mine 13
POMEROY Mountain 45
POORMAN Mine 1 11 13
POPLAR Gulch 47
PORTER & BROWN 201
PORTER Cattle Co. 252
PORTLAND Mine /Mining Co. 301 304
POTOSI Mine 5 13
POTTS, R.F. 73
POUGHKEEPSIE Camp 243
POVERTY Flats 176
POWELL, George 79 Oscar 230
PRESIDENT Mine 178
PRESTON 262 280
PRIDE of SAN JUAN Mine 241
PRINCESS ALICE Mine 61
PRINGLE, Jas. 92
PRIZE Mine 107
PROF. BAUGH'S Reduction Works 91
PROTEAU, Jos. 57
PUEBLO Smelting Co. 192
PUZZLER Mine 19

INDEX

-Q-
QUACKENBUSH, G.W. 273
QUERIDA 86 87
QUEEN of the FOREST Mine 276
QUEEN of the WEST Mine 256
QUINN, --- 200
-R-
RACHOFSKY, H. 108
RACHOFSKY & LONN 108
RAILROAD Gulch 54 55
RAINBOW Lakes 1 16
RAINBOW Route 197
RALEIGH TWELVE Mine 227
RALSTON, Prof. J.P. 125
RANDLE, John 153
RATHBONE 255 256 257
RATLIFF, J.W. 108
RAVEN Mine 298
RAWLEY Mine 228
RAY, Ed. M. 189
RAYMOND, C.H. 234 Geo. N. 234 S.W. 234 239
RAYMOND Mine 142
RAYNE, Geo. T. 57
RECENE 258 267 268 271
RED CLOUD Mine 184 236 245
RED CROSS Tunnel 13
RED ELEPHANT 77
RED ELEPHANT Mining Co. 77
RED ELEPHANT Mountain 77
RED MOUNTAIN District 158 183 192 196 199
RED MOUNTAIN Pass 194 196
RED MOUNTAIN [Town] 192 194 196 197 198 199
RED STOCKINGS 159
REED, Geo. F. 57
REILLY, A.F. 189
REIMER, J.M. 13
REISON, John 49
RENNINGER, Theodore 187
REQUA, G.S. Mining Co. 299
REVENUE Mine 58 200 201
REVENUE TARIFF Mine 256
REXFORD 281
REXFORD Mine Corp. 281
REYNOLDS, Mrs. James 236 Jim 210 211 John 210 211 212
REYNOLDS Mining 78 225
RICE, James 91 N.S. 285
RICH, John 91
RICHARDS, Geo J. 236 J.C. 108 P.A.103 S. 13
RICHARDSON, --- 200 C.L. 299
RICHTER, F. 246
RIDENOUR & HENRY 189
RIGGENS Mine 28
RINGOLD Bros. 176
RISCHE, August 162 163 168
RISING SUN Mine 107
RITCHIE'S PATCH Camp 32
ROBERT E. LEE Mine 165 166
ROBERTS, --- 236 Jenkin 68
ROBINSON 258 267 268 282 283 284 285
ROBINSON, George 166 282 283 Jack 211 John 192 Thos. N. 299
ROBINSON Consolidated Mining Co. 282 284
ROBINSON Hotel 282 285
ROBINSON Mine 192 283 284
ROCHESTER KING Mine 281
ROCHESTER QUEEN Mine 289
ROCK Mine 161
ROCKDALE 44
ROCKVALE 57
RODDA, Thomas 79 William 79
RODMAN, Julius
ROGERS, Buck 94 95 96 Mrs. Mary 299
ROLLER Mine 141
ROMLEY 45 46 47
ROOS Bros. 195
ROOSEVELT, Theodore 134

INDEX

ROOSTER Mine 102
ROPELL, P.F. 135
RORABAUGH 55
ROSITA 26 85 86 88 89 90 91 92 93 255
ROSITA Reduction Works 91
ROUTTE, Gov. 143
ROWE, Geo. 108
ROWSE, Charles 73
ROY, J.B. 73
ROYAL 26 John 45
ROYAL TIGER Mines Corp. 293
RUBY 121 130 131 134
RUBY Camp 131
RUBY CHIEF Mine 133 134
RUBY KING Mine 134
RUBY Mining District 133
RUBY TRUST Mine 200 202
RUBY-ANTHRACITE 121
RUMERY, N.E. 199
RUSSELL 83
RUSSELL Gulch 159
RYAN House 182

-S-

ST. ELMO 25 29 39 45 47 48 49 50
ST. JACOBS Mine 148
ST. JAMES Hotel 57
ST. JOHN 286 287
ST. JOHN Smelting Works 287
ST. JOHN[S] Mine 148 286
ST. KEVINS 178
ST. KEVINS Mine 178
ST. LOUIS Mine 180
ST. MARY'S Glacier 61
SAMPLING HOUSE Saloon 55
SAMPSON Mine 243
SAN BRUNO Mine 147
SAN JUAN 26
SAN JUAN Mine and Mill 245 246
SAN LUIS VALLEY Land and Mining Co. 232
SANGRE DE CRISTO 83

SANTA CLAUS Mine 281
SANTIAGO Mine 81
SARATOGA Lode 4 192
SARGENT, W. 90
SAVAGE BASIN Camp 253
SAWTELL, W.L. 256
SCHAFER'S CROSSING 212
SCHLOSSER, Pete J. 54 56
SCHLUTER, Earnest 127
SCHMIER, Mis[s] R. 92
SCHNEIDER, G.W. 103
SCHOOLFIELD, W.A. 92
SCHRIVER, Henry 92
SCHULTZ WONDER Mine 102
SCHWABE, J.L. 92
SCOFIELD 118 126 143
SCOFIELD, B.F. 143
SCOFIELD Pass 129 143
SCOOPER Mine 165
SCOTLAND, Rev. 118
SCOTT, --- 28 Bros. 13 Wm. 13
SCOTT & CLOW 103
SEARS, WERLEY & CO. 14
SEBASTROM, Pratt
SEFTON, H.T. 84
SEIDEN'S House 158
SELDEMRIDGE, C.B. 298
SENATOR Mine 89 200
SEVEN SISTERS Mining Properties 78
SEVEN-THIRTY Mine 2 13
SHADBOLT, Geo. 256
SHAFT, Martin 21
SHAMROCK Mine 4 161
SHAPTOWN 176
SHARP, Col. 141
SHAVANO 51 52
SHAW, --- 273
SHEA, SMITH & KIMBALL 5
SHEDDON, J. 282
SHEEHAN, Wm. 199
SHEEP Mountain 117 118 267

INDEX

282
SHEEP Mountain Tunnel Mines 118 119
SHEPERD, --- 269 270
SHERMAN 152 153
SHERMAN House 14 15
SHERMAN Lode 11 13
SHERMAN, M. & M. CO. 153
SHERRY'S Saloon 188
SHERWOOD, C.A. 13 Gen. 152
SHIDELER & MORROW 124
SHIRLEY Mine 227
SHOCK, John 262
SHOO FLY Mine 28
SHORT, Mr. and Mrs. 234 Robert 33
SHUEY, Charles 118
SILENT FRIEND Mine 42
SILVER BELL 192
SILVER CITY 64
SILVER CLIFF 86 92 303
SILVER COIN Mine 240
SILVER CREEK 5 78 79
SILVER DOLLAR Saloon 294
SILVER GATE 134
SILVER MOUNTAIN 260
SILVER MOUNTAIN Mining Co. 57
SILVER PLUME 81
SILVER POINT Mine 11
SILVER TROWEL Mine 139
SILVERTON 197 201 236 237 241 243 245
SILVERWING Mine 241
SIMMONS, John 13 Reuben 92
SING LEE 91
SINGLETERRY, Owen 211
SIPPLE Bros. 270
SLABTOWN 162
SLATE Mountain 94 95 96
SLAVICK, Louis 90 91
SMALL, Mrs. M.J. 130

SMELTZER, Nelly 229
SMILE of FORTUNE Mine 152
SMITH, B.F. 91 E.P. 90 91 "General" Jack 295 H.M. 186 James J. 222 Lee 3 Leon 14 Walter H. 3 14
SMITH & COOPER 276
SMITH & CROFT 222
SMUGGLER Mine 253 282
SMUGGLER-UNION Mill 250 Mine 250
SMYTHE, R.K. 270
SNEFFELS 200 201 202
SNEFFELS Mining District 200
SNELL, James 68
SNIVELY Hotel 259
SNOWBLIND Gulch 138
SNOWSLIDE House 256
SNOWSTORM Mine 184
SNOWY RANGE Hotel 88 91
SNYDER, H.C. & CO. 118
SOMERSET 122
SONATA Mine 191
"SONOFABITCH BASIN" 143
SORTER & CO. 90
SOULE, Henry 243
SOUP BONE Musical Club 138
SOUTH LONDON Mine 218
SOUTH MOUNTAIN 224
SOUTH PARK CITY 210
SOVEREIGN PEOPLE Mine 2 11
SOWBELLY Gulch 178
SPALDING & CO. 152
SPANISH BAR 80
SPANISH Camp 232
SPANISH Gulch 83
SPAR Mine 188
SPARKS, John 108
SPENCER Mine 2
SPOTTSWOOD, Mr. 3
SPRING Gulch 70
SPROUT, G.E. 127

INDEX

SQUAW Gulch 295 296
STAATS, --- 269 270
STAINSBY, --- 269
STAINSBY, FORT and COLBY 269 270
STANLEY, Misses 234 J.L. 234 236
STANLEY Mine 48 80
STANSELL, J.B. 207
STAR, S.M. & CO. 201
STAR Mine 120
STAR MOUNTAIN Mine 157
STARK, Tom 49
STARK & GLICK 127
STEDMAN, M. 108
STEIN, Brothers 238 Mr. and Mrs. Frank 234 Mr. and Mrs. Wm. 234
STEPHENS, John 68 William 68
STERLING 154
STEVENS, W.H. 176 Wallace 112 William 161
STEVENS' Lode 81
STEWART Gulch 5
STEWART, --- 88 89 J.J. 49
STEWART Reducing Co. 76
STOCK, John 14
STODDARD, Dr. O.R. 127
STONE, Geo. A. 118
STONE Mine 161 Mining Co.118
STONEWALL 30 53
STONEWALL Mine 30 53
STORER, W.W. 153
STORY, Lt. Gov. 192 S.C. 229
STOWE, Jake 211 212
STRATTON 303 304
STRATTON, Winfred Scott 171 172 303 304
STRATTON Mining Co. 67
STRAY HOUSE Gulch 156 161 172 174
STRINGTOWN 179

STROM, Charles 73 John 68
STRONG, Mrs. 269
STUMPTOWN 180
STUMPH, Joseph 180
SULLIVAN, Mrs. Orion 59
SULTANA Milling Co. 141
SUMMIT HOUSE 269 270 272 291
SUMMIT Mine 197
SUMMIT Tunnel 13
SUMMITVILLE 223 224 225 226
"SUMMITVILLE BOULDER" 225 226
SUMMITVILLE Consolidated Mining Corp. 225 226
SUNDERBERG Mine 107
SUNNYSIDE 191
SUNNYSIDE Mill 242
SUNNYSIDE Mine 188 191 241
SUNSHINE Peak 152
SURLES Mine 280
SUTTON, Marshal 268 S.A. 130
SWAMP ANGEL Mine 184
SWAN [CITY] 288 289 290
SWANDYKE 291
SWANDYKE Gold Mining and Milling Co. 291
SWANSON, F.W. 223
SWEET HOME Mine 204
SWEET WILLIAM Mine 281
SWILL TOWN 175
SWISS BOY Mine 58
SWORDS, Mrs. M. 189
SYLVANITE Mine 123
SYMONS, P.H. 59

-T-

TABOR, Augusta 32 159 160 161 162 163 166 167 169 170 Baby Doe 166 167 169 170 173 174 304 Elizabeth Bonduel Lillie ("Golden Eagle") 171 172 Horace (H.A.W) 28 32 159 160 161 162 163 164 165 166 167

INDEX

168 169 170 171 172 181 207
304 Maxcy 163 166 167 170 172
Rose Mary Echo Silver Dollar
171 172 173
TABOR CITY 169 181
TALLMAN, C.G. 19
TAM O'SHANTER Mine 165
TARRYALL 210 214 220
TARRYALL Creek 214
TARRYALL DIGGINGS 220
TAYLOR, --- 62 Alice 62 Col. 181
 Frank S. 221 John 162 Prof. 236
 S. 214
TAYLOR CITY 181
TAYLOR PARK 131
TEATS, Eugene 38
TEITSWORTH, Rev. 90
TELLER HOUSE 12
TELLURIUM 154
TEN-FORTY Mine 4 13
TEN MILE CITY 282
TEN MILE Mining District 258 267
TENBROOK Mine 185
TERRELL, J.K. 189
TERRIBLE Mine 85
TERRY, Judge John 241
TETON Camp 232
THALER, Frank 234
THATCHER, J.A. 201 M.D. 201
THATCHER Bros. Miners and
 Merchants Bank 201
THIRST PARLOR Saloon 294
THOMAS, C.J. 73 Gov. Charles S.
 151 G.B. 199 G.C. 73 T.S. 73
THOMASSON, --- 90 Tower 91
THOMPSON, G.R. 273 Gassy 256
 257 Pete 4 W.A. 273
THOR Mine 148
THORNTON, Alexander 90
THORP, Louis 14
THREE KINGS Mine 291
TIGER 293

TIGER Lode 293
TILDEN Mine 25 27 28
TIMNEY & KOUTS 229
TIMOTHY Mine 280
TIN CUP 130 303
TIN CUP Gold and Dredging Co. 39
TIP TOP Mine 98 102
TIP TOP Restaurant 238
TOBASCO Mill 154
TOBASCO Mine 154
TOBIE, Wm. 22
TOBIN, Tom 203
TONKIN, Samuel 68
TOLL, Mrs. Charles B. 111
TOLLAND 111
TOLLE, Miss 91
TOMBOY 253 254
TOMBOY Gold Mining Co.
 253 Mine 253 254
TOMICHI 138
TONKYN, Ed. 246
TOPEKA Mine 28
TORPEDO ECLIPSE Mine and
 Mill 200
TOWNSEND, O.P. 90
TRACY, W.C. 273
TRANESCH, George 79
TRAYLOR, Eddy 68 William 68
TREASURE VAULT Mill 99
TREASURY PEAK 117
TREMBATH, Jas. 108
TREMBOTH, Harry 74
TRENOWITH, C. 108
TREVILLION, Wm. 67
TREZONA, Dick 234
TRICKEY, M.M. 246
TRIMBLE, George 166
TROJAN Mine 2
TRUCKEE Lode 118
TRUE, H.A. 19
TRULL, J. 256
TUCKER, H. 90 91

INDEX

TURNER, John 139
TURRELL, Mr. 123
TURRET 54 55 56
TURRET Hotel and Drug Store 54
TWAIN, Mark 36
TWIN LAKES 182 183 196
TWIN LAKES Resort Co. 182
TWISTELTON, William 68
-U-
UNCOMPAHGRE Toll Road 150
UNDINE Mine 145 205
UNION Hotel 270
UNION Mine 75
UTE-ULAY Mine 150 151
-V-
VAN NORDEN, S.E. 190 Mrs. S.E. 190
VANDERVELT Mine 192
VEIS, V.F. 271
VERMILLION Mine 245
VICKSBURG 31 44 57 58 59
VICKSBURG House 57
VICTOR 297 298 301 303
VICTOR Mine 294
VINCENT, T.W. 190
VINCENT'S Opera House 190
VINDICATOR Mine 301 302
VIVANDIERE Barn 54
VIVANDIERE Mine 54
VIRGINIA CITY 130 131
VIRGINIA HILLERTON and ROARING FORKS Toll Road Co. 131
VIRGINIA Mine 13 Smelter 130
VIRGINIUS Mine 200 201
VON BRANDIS, C. 74
-W-
WABASH Mine 13
WAGER Gulch 148
WAGNER, John 248
WAGON WHEEL GAP 236
WAIT, Lew 128

WAIT and PATTERSON 127
WAITE, O.S. 199
WALDORF 81
WALDORF Mining Co. 81
WALKER, John 270 Wm. 285
WALSH, Evalyn 237 Mrs. John 190 Thomas 178 237
WALTERS, T.M. 234
WALTERS & LEE 186
WARD 17 19
WARD, Calvin W. 17 Peter 84
WARNER, J.E. 299 J.W. 90
WARRIN & COULSON 190
WARRIOR'S MARK Mine 264
WASHBURN, E.A. 57
WATERFALL Gulch 250
WATSON, James 20 Mr. 21
WEAVER Brothers 293
WESLEY, Wm. 269 270
WESTERN Hotel 270
WEBB, N.H. 91
WEBB & TOMPKINS 90 91
WEBSTER 221 222
WEBSTER, William Emerson 221
WEBSTER House 222
WEBSTER Pass 221 222
WEINBERGER, L. 84
WEINMERSTADT, E. 270
WEINRICH, W. 190
WEISS, Bros. 225 Henry 279
WELCH Bros. 231
WELLS, -- 160 E.E. 190 Tom 161
WELSH, J.H. 20
WENDLER, Chas. 271 272
WERTLEY, P.J. 13
WESTERN House 269 272
WESTFALL, W.J. 231
WESTOVER, Mr. 5
WETSTEIN Mine 102
WHALEN, Maurice 79
WHAT IS IT Mine 130
WHEEL of FORTUNE Mine 165

INDEX

166 200 228
WHEELER Hotel 276
WHIPP & SAMSEL 135
WHIPPLE Mine 28
WHITE, James 14 Patrick 79
WHITE HOUSE Mountain Quarry 136
WHITE PINE 40 138
WHITE SWAN Mine 281 Mining Co. 291
WHITECROSS 145 154 155 233
WHITEHEAD, J.H. 190
WHITMAN & MOAK 124
WHITNEY, S.W. 273
WHITTENBERGER 48
WIDE AWAKE 112
WIDITCH and CO. 256
WIEDEMANN, Robert 234
WIGHTMAN, James 224 William 224
WIGHTMAN Gulch 224
WILBER, E. 36
WILCOX, Edward John 81
WILCOX & LAKE 84
WILD HORSE Mine 297
WILDEY, A.R. 273
WILEY, Robt. 90
WILKINS, H.T. 269 270
WILLIAMSON, George D. 269 270
WILSEY & RAY 28
WILLIAMS, 74 Mrs. 143 Dan 49 E.C. 68 Frank 92 "Parson" Bill 220 Richard 68 Thomas 68 W.H. 68 William 68
WILLIAMSON, George D. 269 270
WILSON, J.A. 91 James 187 W.D. 103
WINDSOR Hotel 270
WINKLEMAN, --- 271
WINNIE Mine 180
WINTERS, W.K. 186
WOLFE, Joe "Arizona Charlie" 300
WOLFE & SUTTON 84

WOMACK, Bob 303
WOOD, S. 177
WOODRUFF, C.H. 153
WOODS, A.B. 161
WOODSTOCK 113
WILSEY & RAY 28
WINDSOR HOUSE Hotel 88
WINFIELD 31 44 57 58 60
WITTEN, Anna 23 E.A. 23 N.A. 23
WOLF & WEST 77
WOODWARD, Luke 57
WOODS, Wm. H. 13 14
WORK, Henry J. 90
WRAY, Dr. F.H.
WRIGHT, A. 238 Dr. A.E. 28 45 Abe 237 George 200 J.A. 234
WRIGHT & CARLSTROM 238
WULSTEN, Carl 91

-Y-

YANKEE 61 82
YANKEE BOY Mine 28 200
YANKEE DOODLE Mine 161
YANKEE GIRL Mine 192 197 198
YANKEE HILL 61 82
YANKEE Mine 161
YANKTON Mine 245
YELLOW JACKET Mine 191 241
YELLOW MEDICINE Mine 102 146
YOCKEY, L. 49
YOUMAN & CO 91
YOUNG, Wm. 236
YULE Creek 136
YULE Mountain Quarry 136

-Z-

ZIMMERMAN, F. 231

ABOUT THE AUTHORS

Laurel Michele Wickersham and her mother Rawlene LeBaron have long been interested in the turbulent, exciting early history of Colorado. As they hiked across the mountains and researched old newspaper files, city directories, and census records, their findings and notes grew into a book. Researching and writing this book has been a family activity that combined their passion for history with hiking and writing.

Previously, they privately published *The Bonds of Maryland* and *The Missing LeBaron*, two family histories.

Rawlene LeBaron has a BA from Alma College, an MS from Central Michigan University, and has taken additional graduate coursework at The University of Denver and The University of Colorado. She was a copy editor for a major metropolitan newspaper, worked in public accounting, and was a manager in high tech and telecommunications. She is an active CPA in the State of Washington. She is a member of Mensa and Intertel.

Laurel Michele Wickersham is an artist and trained as a paralegal before focusing on history.

They are members of numerous historical and genealogical societies, including the Boulder Historical Society, Colorado Historical Society, Society of Indiana Pioneers, The Winthrop Society, Society of Magna Charta Dames, The Plantagenet Society, and the Colonial Order of the Crown.